DELTA II & III SPACE OPERATIONS AT CAPE CANAVERAL

1989 – 2009

by Mark C. Cleary

45th SPACE WING
History Office

PREFACE

This study addresses DELTA II and DELTA III space operations at Cape Canaveral, Florida, from the DELTA II's first launch in 1989 through the DELTA II's most recent launch (as of this writing) in late March 2009. The 45th Space Wing has been – and continues to be – responsible for ensuring public safety for all space-related operations on the Eastern Range. The Wing's 45th Launch Group and its 5th Space Launch Squadron (5 SLS) continue to provide government oversight for DELTA IV and ATLAS V space launch operations at the Cape. Those units are ably supported by the Launch Group's 45th Launch Support Squadron (45 LCSS) as well as the 45th Operations Group's 45th Operations Support Squadron (45 OSS) and 1st Range Operations Squadron (1 ROPS), not to mention a whole host of other agencies with Range-related missions.

Nevertheless, the inactivation of the 1st Space Launch Squadron (1 SLS) and the anticipated transfer of DELTA II facilities to NASA in late 2009 mark the end of an Air Force effort dating back to the first THOR missile launch in late January 1957. To commemorate this turn of events, I re-edited an earlier work "DELTA Space Operations at the Cape, 1993-2001." I also used excerpts from "The Cape: Military Space Operations, 1971-1992," "The 6555th: Missile and Space Launches Through 1970," and the 45th Space Wing annual histories I have written since 2001.

The study provides the reader with a basic summary of DELTA II- and DELTA III-related programs, changes in organization, and improvements in Complex 17's launch facilities. It also presents military, civil service and contractor roles in what is decidedly a „team' effort. The lion's share of the material relates to various launch campaigns and individual flights. The DELTA III flights are grouped all together in Chapter IV, but the DELTA II missions are placed in three separate categories based on their customers: 1) military, 2) civilian agency, and 3) commercial satellite service providers. All DELTA customers rely on commercial launch contractors to put their respective payloads into orbit. So, for our purposes, the type of agency sponsoring a payload (the U.S. Air Force, the United Kingdom's Ministry of Defence, NATO, NASA or a commercial satellite service company) determines where its mission summary is placed.

Range officials mark all launch times in Greenwich Mean Time, as indicated by a "Z" at various points in the narrative. Unfortunately, that convention creates a one-day discrepancy between the local date (reported in the media) and the "Z" time's date whenever the launch occurs late at night but before midnight. Competent authorities have reviewed all the material presented in this study, and it is releasable to the general public. Any factual errors noted by the reader may be addressed to the 45th Space Wing History Office, 1201 Edward H. White II Street, Patrick AFB, FL 32925-3299.

MARK C. CLEARY

June 2009

TABLE OF CONTENTS

(Blank Page)

CHAPTER I

ORIGINS

Following the launch of America's first satellite at the end of January 1958, U.S. space technology matured rapidly in the 1960s and 1970s. In large part, this was due to our competition with the Soviet Union, though the commercial and scientific benefits of space exploration spurred progress as well. Defense Department officials soon found many ways to apply space systems to the nation's weather, communications, navigation, surveillance and early warning missions. Air Force-led space operations became an indispensable part of the overall U.S. defense effort by the late 1970s. Unfortunately, space systems management was shared among three very different organizations – Air Force Systems Command, North American Aerospace Defense Command, and Strategic Air Command. To remedy that fragmentation of its space effort, the Air Staff decided to place space systems operations under a separate major command and create an organization within the Air Staff to supervise the effort. The Directorate of Space Operations was set up under the Air Force's Deputy Chief of Staff for Operations in October 1981, and the Air Force created Air Force Space Command (AFSPACECOM) on 1 September 1982.[1]

AFSPACECOM eventually gained operational control of the Satellite Early Warning System, the Defense Meteorological Satellite Program (DMSP), the NAVSTAR Global Positioning System (GPS), and various ground surveillance, satellite communications, and control systems. In the meantime, Air Force Systems Command (AFSC) had its Space Division procure launch vehicles, upper stages and spacecraft for the Air Force even after control of the various satellite systems passed to AFSPACECOM. Procurement responsibilities were subsequently transferred to the Space and Missile Systems Center (SMC) under Air Force Materiel Command in the early 1990s.[2]

From the organizational perspective, the transfer of space launch operations from AFSC to Air Force Space Command was AFSPACECOM's most significant event in the early 1990s. At long last, space operations – from lift-off to satellite deactivation – were assigned to a major *operational* command. Though the transition to full operational status took several years, the path was laid out clearly in AFSPACECOM's Programming Plan 90-2, which was signed by General Ronald W. Yates (for AFSC) and Lt. General Thomas S. Moorman, Jr., (for AFSPACECOM) at the end of August 1990.[3]

According to that plan, Phase I of the Launch Transfer began on 1 October 1990. On that date, the Eastern Space and Missile Center (ESMC), the Western Space and Missile Center (WSMC), and their associated range organizations were transferred from AFSC to AFSPACECOM along with the 6550th Air Base Group, Patrick Air Force Base, Cape Canaveral Air Force Station, and AFSC Hospital Patrick. The 6555th and 6595th Aerospace Test Groups on the east and west coasts of the United States remained with AFSC, but two new organizations –

---

[1] Space Division History, FY 1982, Vol I, pp 12, 13, 14; AFSC History, FY 1983, Vol I, pp 21, 22.

[2] AFSPACECOM History, CY 1990, pp 1, 2; 45 SW History Office, "The Cape: Military Space Operations, 1971-1992," Jan 1994, pp 20, 21; Discussion, M. Cleary with Dr. Harry N. Waldron, SMC Historian, 22 Apr 2009.

[3] ESMC History, FY 1990, Vol I, pp 383, 384; 9th Space Division History, FY 1991, p 3; Headquarters AFSPACECOM Programming Plan 90-2, 15 Aug 1990, p 3; 45 SW History, 1 Oct 1990 - 31 Dec 1991, Vol I, pp 39, 40.

the 1st and 2nd Space Launch Squadrons – were activated and assigned to AFSPACECOM under ESMC and WSMC respectively on 1 October 1990.[4]

The 1st Space Launch Squadron (1 SLS) was constituted from resources taken from the 6555th Aerospace Test Group at Cape Canaveral Air Force Station. It became AFSPACECOM's DELTA II launch squadron. The 2nd Space Launch Squadron (2 SLS), which was created from resources taken from the 6595th Aerospace Test Group at Vandenberg AFB, became an ATLAS E squadron under WSMC. Though the 6555th Aerospace Test Group's ATLAS and TITAN resources were not ready to become operational squadrons on 1 October 1990, Air Force Systems Command transformed them into ATLAS II and TITAN IV Combined Test Forces (CTFs) to serve both major commands until such time as they could become operational squadrons under AFSPACECOM.[5]

Following the inactivation of ESMC and the activation of 45th Space Wing on 12 November 1991, the 1 SLS was reassigned from ESMC to the 45th Operations Group. Both CTFs were placed under the 45th Operations Support Squadron of the 45th Operations Group until they could become fully operational squadrons in their own right. Most of the people assigned to the new organizations performed many of the tasks they handled before the transfer, but their training and reporting procedures became increasingly operational in nature. The ATLAS II CTF became the 3rd Space Launch Squadron (3 SLS) after the second military ATLAS II/CENTAUR launch on 2 July 1992. The TITAN IV CTF was reassigned directly under the 45th Operations Group in the summer of 1992. It was activated as the 5th Space Launch Squadron (5 SLS) on 14 April 1994. The 6555th Aerospace Test Group was deactivated when AFSC and Air Force Logistics Command merged to become Air Force Materiel Command on 1 July 1992.[6]

Complex 17 and the Old DELTA Program

Launch Complex 17 was built for the Air Force's THOR Intermediate Range Ballistic Missile (IRBM) research and development program in 1956. The site consisted of two launch pads (17A and 17B) and one blockhouse. In addition to providing launch facilities for THOR IRBM test flights through mid-December 1959, Complex 17 began supporting space launches in the late 1950s. The site was modified in the early 1960s to support a whole host of launch vehicles derived from the basic THOR missile booster. The most common variant – called the DELTA – was used to launch NASA payloads, but the Air Force also launched six THORs on Aerothermodynamic/Elastic Structural Systems Environmental Test (ASSET) missions from Complex 17 between 18 September 1963 and 24 February 1965. Having no further immediate need for the site, the Air Force transferred Complex 17 to NASA in the spring of 1965. The complex continued to support DELTA launch operations under NASA's sponsorship until it was

[4] Headquarters AFSPACECOM Programming Plan 90-2, 15 Aug 1990, p 3; 45 SW History, 1 Oct 1990 - 31 Dec 1991, Vol I, pp 39, 40.

[5] ESMC/45 SW History, 1 Oct 1990 - 31 Dec 1991, Vol I, pp 39, 40; "Second Atlas II launch activates new squadron," *45th Space Wing Missileer*, 10 Jul 1992; Item, "Group deactivates after merger," *45th Space Wing Missileer*, 10 Jul 1992; Interview, M. Cleary with Mr. Jeffrey Geiger, 30th Space Wing Historian, 7 Jul 1993.

[6] ESMC/45 SW History, 1 Oct 1990 - 31 Dec 1991, Vol I, pp 43, 44; 45 SW History, CY 1992, Vol I, p 15; Item, "Group deactivates after merger," *45th Space Wing Missileer*, 10 Jul 1992; HQ AFSPC Special Order GD-016, 4 Apr 1994; "5th SLS joins wing," *45th Space Wing Missileer*, 15 Apr 1994.

returned to the Air Force in October 1988 to support the newly emerging DELTA II launch vehicle program.[7]

The DELTA space launch vehicle program was transferred from NASA to the Air Force pursuant to a Memorandum of Agreement (MOA) signed by NASA Administrator Dr. James C. Fletcher and Air Force Secretary Edward C. Aldridge, Jr. in June and July 1988 respectively. To complete the orderly transfer of DELTA launch site operations at the Cape, Colonel Lawrence L. Gooch (ESMC Commander) and KSC Director Forrest S. McCartney (Lt. General, USAF Ret.) signed a Memorandum of Understanding (MOU) in mid-August 1988. NASA sponsored its final DELTA payload processing operation in 1989, but all DELTA space launches from Complex 17 were actually under Air Force supervision from mid-October 1988 onward.[8] The very first DELTA II flight was the NAVSTAR Global Positioning Satellite (GPS II-1) mission. It was launched from Pad 17A on 14 February 1989.[9]

Between mid-October 1988 and 1 October 1990 – the period just before the 1 SLS was activated – three DELTA missions were launched from Complex 17. All of them lifted off Pad 17B. The first was DELTA 183, an experimental Strategic Defense Initiative (SDI) mission.[10] A countdown for DELTA 183 was scrubbed on 15 March 1989 due to vehicle and spacecraft problems, but the next countdown on 24 March 1989 went smoothly, and the mission was successful. The second mission featured a British commercial communications satellite, the BSB-R1. It was launched on 27 August 1989, and it had the distinction of becoming the first commercial payload ever launched into orbit by a U.S. commercial Expendable Launch Vehicle (ELV). The third launch supported the INSAT-1D, a multi-purpose communications satellite developed by Western Laboratories (Ford Aerospace) for India's Department of Space. The launch vehicle was a DELTA Model 4925, and it placed the satellite in its pre-planned 90 x 21,321-nautical-mile orbit on 12 June 1990. The INSAT-1D was the last of the old DELTA missions launched from Complex 17.[11]

---

[7] Mark C. Cleary, 45th Space Wing History Office, "The 45th Space Wing: Its Heritage, History and Honors, 1950-2007," updated 19 May 2008.

[8] In a letter to ESMC dated 18 August 1988, Mr. James D. Phillips detailed the transfer of accountability for Area 57, most of Hangar K, and all of Complex 17 – except for NASA office spaces – from NASA to the Air Force. NASA also released Area 55 facilities that it had on loan from the Air Force. Complex 17's real property was in Air Force hands by 10 October 1988.

[9] "Memorandum of Agreement between the United States Air Force (USAF) and the National Aeronautics and Space Administration (NASA) for Transition of the Delta Space Launch Vehicle Program," signed 13 June and 1 July 1988; "Memorandum of Understanding between the John F. Kennedy Space Center (KSC) and the Eastern Space and Missile Center (ESMC) for Transition of the Delta Launch Site Operations at Cape Canaveral Air Force Station (CCAFS) from NASA to USAF," signed 15 and 16 August 1988; ESMC History, FY 1989, Vol I, p 162; Letter, Mr. James D. Phillips, KSC/DF-ESD-2, to ESMC/DER, "Transfer of Real Property (Delta Program)," 18 Aug 1988.

[10] In a speech given in March 1983, President Ronald Reagan proposed the Strategic Defense Initiative (SDI) as a means of rendering nuclear weapons obsolete. He later redefined SDI as a research program utilizing new technologies to create effective defenses against ballistic missiles. In support of the President's proposal, the Defense Department established the Strategic Defense Initiative Organization in late March 1984. The majority of the effort was undertaken by the Air Force and Army. The Air Force managed efforts related to boost and mid-course phases of ballistic missile interdiction; the Army managed terminal phase/ground defense programs.

[11] 45 SW History Office, "The Cape: Military Space Operations, 1971-1992," Jan 1994, pp 21, 22; ESMC History, FY 1989, Vol I, pp 366, 369; ESMC History, FY 1990, Vol I, p 318.

The NAVSTAR Global   Positioning System and the DELTA II Launch Debut

The NAVSTAR GPS program opened up a new field for space support operations at the Cape in the 1980s: the launching of satellites to provide highly accurate three-dimensional ground, sea, and air navigation. The U.S. Navy and U.S. Air Force began the effort in the early 1960s with studies and experiments to test the feasibility of using satellite-generated radio signals to improve military navigation. After 10 years of extensive research, the services concluded that Defense Department requirements would be served best by a single, highly precise, satellite-based Global Positioning System (GPS). The technology necessary to field the GPS was confirmed, and a total of six Block I NAVSTAR satellites were launched on ATLAS F boosters from Vandenberg's Space Launch Complex 3 (East) between 22 February 1978 and 27 April 1980.[12]

By the end of 1980, the NAVSTAR GPS constellation was arranged in two orbital planes of three satellites each, orbiting Earth at an altitude of approximately 10,900 nautical miles. Five more Block I satellites were launched from Vandenberg AFB between 18 December 1981 and 9 October 1985, and four of those missions were successful. Following up on the GPS' initial success, the Air Force planned to procure and deploy a constellation of 24 Block II GPS satellites via the Space Shuttle by the end of 1987. Unfortunately, funding cuts in 1980 and 1981 reduced the planned constellation to 18 Block II satellites. The cuts also added a year to the Block II deployments.[13]

Following the Space Shuttle *Challenger* disaster in January 1986, the GPS Program Office diverted the first eight Block II GPS satellites from the Shuttle program to a brand new "Medium Launch Vehicle" later known as the DELTA II. The Shuttle was still being considered as a launch vehicle for later GPS missions, but its prospects dwindled under increased pressure to replenish the GPS constellation to meet user needs quickly. Only two NAVSTAR II satellites remained on the Shuttle's manifest by the end of FY 1988, and they were reassigned to the DELTA II program in 1989.[14]

Space Division awarded the Medium Launch Vehicle (MLV) contract to McDonnell Douglas Astronautics Company on 21 January 1987. Together with its options, that DELTA II contract was valued at $669,000,000. McDonnell Douglas also had at least four firm orders from non-military customers to launch DELTA II vehicles on commercial missions. However, unlike earlier commercial arrangements, the company would no longer be under contract to NASA. Under the new Commercial Expendable Launch Vehicle program encouraged by President Reagan since 1983, McDonnell Douglas would be responsible for producing, marketing and launching its commercial DELTA IIs. The Air Force would be responsible for ensuring safety and environmental standards for commercial as well as military launches, but McDonnell Douglas would have greater responsibility in meeting those standards – including quality control. Launch Pads 17A and 17B would be equipped to handle commercial as well as Defense Department

[12] AFSC History, FY 1980, Vol I, pp 531, 533, 534; Jeffrey Geiger, 30th Space Wing History Office, "Launch Summary," 15 Dec 1992, pp 83, 84, 85, 87; SAMSO History, CY 1978, Vol I, p 63; SAMTO and WSMC History, CY 1980, Vol I, pp 52, 54.

[13] Space Division History, FY 1981, Vol I, pp 246, 247; Space Division History, FY 1982, Vol I, pp 192, 193; Space Division History, FY 1983, Vol I, pp 176, 178; Space Division History, FY 1984, Vol I, pp 180, 186; SAMTO and WSMC History, FY 1986, Vol I, p 66; Geiger, "Launch Summary," pp 89, 91, 92, 93, 94.

[14] Space Division History, FY 1986, Vol I, pp 252, 253, 254; Space Division History, FY 1987, Vol I, pp 196, 197; Space Systems Division History, FY 1988, Vol I, pp 306, 309, 315; Space Systems Division History, FY 1989, Vol I, p 342

missions. McDonnell Douglas and its subcontractors were soon hard at work preparing the pads for the new DELTA II vehicles.[15]

The DELTA IIs would be taller than earlier DELTA vehicles, so one of the contractors' first tasks was to raise Complex 17's Mobile Service Towers (MSTs) 20 feet to accommodate the DELTA II's longer stages. Other modifications revolved around Pad 17A initially because Pad 17B was committed to the DELTA 181 mission which was scheduled to be launched in February 1988. The contractor expanded his workweek to 10 hours per day, seven days a week in January 1988. Pad 17A's modifications were essentially complete by mid-April 1988, and Pad 17B's work was on schedule. The contractor's remarkable progress was due in large part to having most of the offsite prefabrication work completed before modifications on Pad 17B began.[16]

Unfortunately, trouble loomed from a different quarter in July 1988: McDonnell Douglas ran into trouble getting some fiber optic equipment it ordered for Pad 17A. Consequently, the first DELTA II launch was rescheduled from 13 October 1988 to 8 December 1988. Test discrepancies in McDonnell Douglas' plant delayed the first launch somewhat longer, but the first DELTA II's first stage was erected on Pad 17A by 2 November 1988. The vehicle's interstage was installed at the pad on 5 November, and the solid rocket motors were mated to the vehicle a few days later. The range contractor and ESMC's engineers completed launch data connections between the blockhouse and Pad 17A around the middle of November 1988. On 24 January 1989, command and telemetry verification tests confirmed reliable links between Sunnyvale and Colorado Springs for the upcoming NAVSTAR II GPS mission. Following final prelaunch tests, the countdown was picked up on 12 February 1989, but it was scrubbed at 1827Z due to excessive high altitude winds. The countdown was picked up again on 14 February, and the lift-off was recorded at 1829:59.988Z on 14 February 1989. The first DELTA II placed the first NAVSTAR II GPS payload in the proper transfer orbit. The mission was a success.[17]

DELTA II Operations Overview

(Readers please note: This section presents a *very basic* description of the principal components of the DELTA II and how McDonnell Douglas – the original DELTA II contractor – processed launch vehicles in the early 1990s. While some of the principal 45th Space Wing agencies that supervised DELTA II-related operations over the past two decades are presented under "*Military Oversight*" later in this chapter, not every agency is included. The 45th Weather Squadron, the 45th Civil Engineer Squadron, and the 45th Range Management Squadron could be included as supporting agencies, but this study attempts to focus on those squadrons most closely associated with DELTA II launch vehicles and their payloads. The operations were complex, and the various operators' titles, techniques, training, responsibilities and equipment were susceptible to change.)

Though the DELTA II's launch debut in February 1989 drew considerable attention, it should be remembered that the new vehicle — like the basic DELTA space launch vehicle — was rooted in the Air Force's THOR IRBM of the late 1950s. The DELTA II (Model 6925) was somewhat larger and more powerful than the original DELTA, but it was configured in much the same way as its predecessor. Its first stage was eight feet in diameter – the same diameter as the old DELTA's first stage – but it was 12 feet longer (e.g., 85.7 feet long vs. 73.7 feet long). The

---

[15] ESMC History, 1 Oct 1984 - 30 Sep 1986, Vol I, p 39; ESMC History, FY 1988, Vol I, pp 102, 103, 104; ESMC History, FY 1989, Vol I, pp 161, 162.

[16] ESMC History, FY 1988, Vol I, pp 104, 105; ESMC History, FY 1989, Vol I, p 361.

[17] ESMC History, FY 1988, Vol I, p 106; ESMC History, FY 1989, Vol I, pp 161, 167, 168, 364.

first stage was equipped with a Rocketdyne RS-27 main engine, which had been introduced to the DELTA line of vehicles in the late 1970s. The RS-27 was rated at 207,000 pounds of thrust at sea level and 231,700 pounds at altitude. The main engine burned RP-1 (i.e., highly refined kerosene) and liquid oxygen. Nine 36.6-foot-long Thiokol Castor IVA Solid Rocket Motors (SRMs) surrounded the first stage. Each of those motors was rated at 97,000 pounds of thrust at sea level.[18]

When the DELTA II was upgraded in 1990 to lift heavier GPS Block IIA satellites, two important modifications were incorporated into the new (Model 7925) vehicle. First, the expansion ratio on Rocketdyne's main engine nozzle was increased from 8:1 to 12:1 to create an improved version of the RS-27 called the RS-27A. That modification boosted the RS-27A's engine performance from 231,700 pounds to 237,000 pounds of thrust at altitude. Second, more powerful and longer Alliant Techsystems Graphite Epoxy Motors (GEMs) replaced the Thiokol Castor IVA SRMs. Each of the GEMs was 42.5 feet long, and each provided 98,870 pounds of thrust at lift-off.[19]

The first DELTA II (Model 6925) was launched from the Cape on 14 February 1989. The first Model 7925 was launched on 26 November 1990. On both models, six of the DELTA II SRMs[20] fired just before lift-off. The remaining three motors fired approximately one minute after lift-off, just about two seconds after the first six SRMs burned out. Each air-lit Castor IVA solid produced 108,700 pounds of thrust. Each air-lit GEM provided 110,820 pounds of thrust.[21]

The second stage on both DELTA II models was 5.7 feet in diameter and 19.6 feet long. Like the second stages on the old DELTAs, the DELTA II's second stage was equipped with an 8-foot diameter miniskirt to support its engine and propellant tanks and shelter its umbilical interfaces and antennas. The miniskirt was attached to the interstage structure (lower end) and the payload fairing (upper end) to give the DELTA II a "straight eight" profile. The second stage was equipped with an Aerojet AJ10-118K engine that burned Aerozine 50 and nitrogen tetroxide. The second stage delivered approximately 9,645 pounds of thrust. The DELTA II's third stage was a derivative of the PAM-D equipped with a Thiokol Star 48B solid rocket motor. The third stage delivered approximately 15,100 pounds of thrust. Both models of the DELTA II were originally equipped with 9.5-foot diameter fairings, and the Model 7925 could also carry the new 10-foot diameter payload fairing. Both vehicles stood 129.9 feet tall. The Model 6925 could place a 3,190-pound payload into geosynchronous orbit. The Model 7925 could lift a 4,120-pound payload into geosynchronous orbit.[22]

---

[18] Summary, ANSER, "A Historical Look at United States Launch Vehicles, 1967 - Present," 1997, pp III.A-1, III.A-2, III.A-5, III.A-7, III.A-9, III.A-13, III.A-16, III.A-17; Excerpt, McDonnell Douglas, "PRD/OR 5516, KOREASAT 1& 2," Dec 1994, p 1310.1.

[19] 45 SW History, CY 2007, Vol I, pp 89, 90.

[20] The GEMs were also SRMs in the generic sense of the term. In this narrative, the term 'GEMs' is used to identify any solid rocket motors associated with 7xxx series DELTA II launch vehicles.

[21] Program Requirements Document, McDonnell Douglas Space Systems Co., " PRD Revision 1 DELTA II, 6920 Vehicle," Nov 1989, pp 1310.1, 1310.2, 1311.1; Summary, McDonnell Douglas, "Commercial DELTA II Payload Planners Guide," Dec 89, Tables 1-2, 1-5, pp 1-3, 1-4; Briefing Slides, M. L. Baldwin, McDonnell Douglas, "Graphite Epoxy Case Rocket Motor (GEM) for 7925 Delta II Launch Vehicle, Design, Processing and Handling Overview," 29-30 March 1990; Summary, ANSER, "A Historical Look at United States Launch Vehicles, 1967 - Present," 1997, pp III.A-1, III.A-2, III.A-5, III.A-7, III.A-9, III.A-13, III.A-16, III.A-17; Excerpt, McDonnell Douglas, "PRD/OR 5516, KOREASAT 1& 2," Dec 1994, p 1310.1.

[22] Program Requirements Document, McDonnell Douglas Space Systems Co., "PRD Revision 1 DELTA II 6920 Vehicle," Nov 1989, pp 1310.1, 1310.2, 1311.1; Summary, ANSER, "A Historical Look at United States Launch Vehicles, 1967 - Present," 1997, pp III.A-7, III.A-9, III.A-16, III.A-17; Excerpt, McDonnell Douglas, "PRD/OR 5516,

McDonnell Douglas completed an avionics upgrade on the DELTA II launch vehicle in 1995 and introduced a new Advanced Launch System to modernize DELTA II ground processing functions. In May 1995, technicians and engineers erected a DELTA II pathfinder vehicle on Pad 17A to validate both improvements. The 1970s vintage avionics suite – consisting of flight control, power, telemetry, and flight termination systems – was replaced with state-of-the-art electronics. The main components included the Redundant Inertial Flight Control Assembly (RIFCA) guidance and navigation system,[23] the Power and Control Box (used for general control and load control), and the Master Telemetry Unit. Allied Signal in Teterboro, New Jersey, built the RIFCA. Gulton Data Systems in Albuquerque, New Mexico, manufactured the telemetry system. Cincinnati Electronics in Mason, Ohio, supplied the command receiver decoder. McDonnell Douglas built the Power and Control Box.[24]

The Advanced Launch Control System (ALCS) delivered comparable improvements to McDonnell Douglas' ground processing and checkout operations. The ALCS used high-speed commercial computer workstations and fiber optics to display system status graphically on a continuously updated basis. The ALCS automatically analyzed data and reported normal processes along with any anomalies that surfaced during processing.[25] The system used standard interfaces and software, and the contractor could expand it to include expert systems and advanced automation techniques at a later date. Following pathfinder testing, McDonnell Douglas employed the RIFCA and ALCS operationally for the first time on the DELTA II Rossi X-Ray Timing Explorer (RXTE) mission in December 1995. The operation was successful.[26]

In an interesting turn of events, negotiations between the Boeing Company and McDonnell Douglas in 1996 eventually led to the purchase of McDonnell Douglas by Boeing in the summer of 1997. McDonnell Douglas became a wholly owned subsidiary of the Boeing Company on 1 August 1997.[27] The old McDonnell Douglas workforce remained at Cape Canaveral after the deal was concluded, but they were officially employees of Boeing Expendable Launch Services

---

KOREASAT 1& 2," Dec 1994, p 1310.2; Fact Sheet, McDonnell Douglas, "DELTA II Family of Launch Vehicles," Jul 1996; Handout, the Boeing Company, "DELTA III – the Commercial Evolution of DELTA," 3 Apr 1998.

[23] The RIFCA employed redundant laser gyros and accelerometers to sense the DELTA II's velocity, angular position and direction. Data was constantly updated and acted on by three on-board computers to allow the RIFCA to maintain control of the vehicle throughout its flight.

[24] 45 SW History, CY 1995, Vol I, p 109.

[25] Veda Systems in Maryland provided the command and data processing system for the ALCS. Aydin Vector, Newtown, Pennsylvania, supplied the ground data acquisition system.

[26] News Release, McDonnell Douglas Aerospace, "DELTA Launch Vehicle Avionics Upgrade and Advanced Launch Control System Background Information," 10 Jan 1996; "Delta blazes new trail with pathfinder," *45th Space Wing Missileer*, 19 May 1995; "New system replaces avionics on Delta," *45th Space Wing Missileer*, 19 Jan 1996; Summary, McDonnell Douglas Aerospace, "Avionics Upgrade Vehicle Electrical Sequencing Summary," undated.

[27] In his letter to the Honorable Robert Pitofsky, Chairman of the Federal Trade Commission, Deputy Secretary of Defense John P. White advised Mr. Pitofsky on 1 July 1997 that the Department of Defense considered the proposed acquisition of McDonnell Douglas by the Boeing Company "acceptable." The decision followed the Department's comprehensive review of the proposed transaction and its potential effect on the various defense contracts held by both companies. The Defense Department concluded that it would be able to maintain competition for its various purchases after the merger was concluded. The decision and declaration cleared the way for Boeing's purchase of McDonnell Douglas in August 1997.

thereafter. Boeing rightfully took credit for DELTA II launch vehicle operations at the Cape after 1 August 1997.[28]

Then, in May 2005, Boeing and Lockheed Martin announced their plans to form United Launch Alliance (ULA). Despite complaints from the companies' competitors (notably Northrop Grumman), the U.S. Federal Trade Commission approved the joint venture on 3 October 2006. ULA opened its doors for business on 1 December 2006 as a 50-50 partnership for its parent companies. ULA was headquartered in Denver, Colorado, and its major assembly and integration operations for ATLAS V as well as DELTA II and DELTA IV launch vehicles were consolidated at what had been Boeing's DELTA plant in Decatur, Alabama.[29] DELTA II launch contractors were ULA employees as of 1 December 2006, and ULA took credit for any DELTA IIs launched after that date.[30]

With the exception of the NAVSTAR Global Positioning System (GPS) missions that placed spacecraft in any one of six orbital planes 10,898 nautical miles above Earth,[31] the flight profiles for DELTA II missions were as varied as their payloads. On the other hand, the Eastern Range's support requirements for all DELTA II flights were remarkably similar. The support included radar coverage from assets on the Cape, Patrick AFB, Merritt Island, and instrumentation stations on Antigua and (at least for the early flights) Ascension. Cape Canaveral and Jonathan Dickinson Missile Tracking Annex (JDMTA) provided command/destruct coverage for all DELTA II missions, and Antigua provided additional command/destruct functions for missions not involving northerly azimuths.[32] Telemetry systems on Merritt Island supported all DELTA II flights. Additional telemetry support came from New Hampshire on northerly missions. Antigua and Ascension provided telemetry for missions not involving northerly flight azimuths.[33]

Optical support for DELTA II flights in the 1990s included Distant Object Attitude Measuring System (DOAMS), Remote Optical Tracking Instrument (ROTI), and fixed Intercept Ground Optical Recorder (IGOR) assets at Cocoa Beach, Melbourne Beach, and Patrick AFB.

---

[28] News Release, U.S. Department of Defense, "DOD Finds Boeing's Acquisition of McDonnell Douglas Acceptable," No. 351-97, 1 Jul 1997; New Release, Boeing, "McDonnell Douglas Commences a Fixed Spread Tender Offer on Notes," (cited as Boeing subsidiary), 6 Oct 1997.

[29] It should be noted that various components continued to be built elsewhere. For example, ATLAS V mechanical structures, payload fairings, and adapters were built in Harlingen, Texas.

[30] 45 SW History, CY 2006, Vol I, pp 39, 40.

[31] Typically, the DELTA IIs carrying GPS payloads lifted off the launch pad and rolled into a flight azimuth of 110 degrees. As mentioned earlier, six GEMs ignited just before lift-off, and the remaining three GEMs ignited shortly after the first six GEMs burned out. The first six GEMs dropped off the vehicle a couple of seconds later, and the remaining three GEMs continued to thrust until they burned out and separated from the DELTA II about 130 seconds into the flight. The DELTA II first stage engine cut-off occurred about 260 seconds after lift-off. The first stage separated from the vehicle eight seconds later, and the second stage ignited soon after. The payload fairing was jettisoned during the second stage's first burn, which was completed around 11 minutes into the flight. Following a coasting period, the second stage restarted and continued to thrust for about 20 seconds. Then it coasted along until it separated from the launch vehicle. For GPS II and early GPS IIR (Replenishment) satellite missions, the third stage ignited about 22 minutes after lift-off, and it continued thrusting for about 90 seconds. For GPS IIR Modernized payloads, the vehicle coasting period was a lot longer. The third stage ignited more than one hour after liftoff. In either instance the spacecraft separated about two minutes after the third stage shut down, and the payload was injected into its pre-selected orbit.

[32] Wallops Island and Argentia Missile Tracking Annex (Newfoundland) provided command/destruct coverage for northerly missions.

[33] 45 SW History, CY 1993, Vol I, pp 209, 210; 45 SW History, CY 1997, Vol I, pp 92, 93, 101; 45 SW History, CY 2000, Vol I, pp 106, 109; 45 SW History, CY 2001, Vol I, pp 95, 96; 45 SW History, CY 2007, Vol I, pp 91, 92, 99.

Later DELTA flights required the services of two DOAMS at Patrick AFB and Playalinda Beach, two Advanced Transportable Optical Tracking Systems (ATOTS) on Merritt Island, and at least one other optical site on Cape Canaveral or Merritt Island. When computers, communications, weather services, HH-60G surveillance helicopters, a weather reconnaissance Learjet, and off-range resources were added to the support picture, it is easy to see why every DELTA II launch was a major effort.[34]

Like the TITAN and ATLAS lines of vehicles, the DELTA II line was built with major components supplied by several different contractors. As operations got underway in the late 1980s, McDonnell Douglas built the basic core vehicle and supplied fairing materials at its plant in Huntington Beach, California. Then it shipped those items to another plant in Pueblo, Colorado, for further assembly and/or match ups with other contractors' components. Rocketdyne provided the DELTA II's main engine, and Aerojet supplied the vehicle's second stage engine. DELCO supplied the inertial guidance system, and Morton Thiokol built the strap-on solid rocket motors used for the basic Model 6925 DELTA II vehicle. (As mentioned earlier, Alliant Techsystems provided the more powerful GEMs used on the Model 7925.) Once the core vehicle stages and fairing were assembled, they were transported by truck to the Cape and received at Hangar M. After their initial inspection at Hangar M, the first and second stages were taken to the DELTA Mission Checkout Operations Area where they received telemetry and controls checks, flight simulations, and dual composite testing. The first and second stages were transported to the Horizontal Processing Facility (HPF) in Area 55 for destruct system installation. Following processing at the HPF, both stages were moved to Complex 17 and erected.[35]

The strap-on solids and the new DELTA-configured PAM-D followed different processing flows. The DELTA II's Morton Thiokol or Alliant Techsystems SRMs were trucked to the Cape's Area 57 initially. There the solids were: 1) inspected, 2) checked for leaks and flaws in the solid propellant, and 3) built up with the required destruct harnesses and nose cones. Then the assembled solid rocket motors were placed on transporters and moved out to Complex 17. The DELTA II's PAM-D motor was received at an ordnance storage area where it was inspected, placed in a cold chamber, and cold-soaked. Later the motor was transferred to the Non-Destructive Test Laboratory (NDTL) where it was x-rayed and placed in a shipping container for transport to the NAVSTAR Processing Facility (NPF). At the NPF the motor was assembled into a complete PAM-D by adding a payload attachment fitting equipped with a fueled nutation control system and a spin table. Following assembly, the PAM-D was spin-balanced at NASA's Explosive Safe Area 60 located on the Cape. Then the PAM-D was returned to the NPF and mated to the spacecraft. The payload was installed in a McDonnell-Douglas payload container, loaded on a transporter, and driven out to the launch pad.[36]

At Complex 17 the entire process came together to create a complete DELTA II launch vehicle. The interstage and payload fairing were brought out to the launch pad from Hangar M. The first stage, solid rocket motors, interstage and second stage were erected and mated at the pad, and the payload fairing was secured in the Mobile Service Tower (MST). Technicians aligned the solid rocket motors, and umbilicals were installed. Electrical and mechanical qualification checks were accomplished about a month before the launch, and subsystem checks continued as the contractor prepared to mate the vehicle with its payload about nine days before lift-off. Ordnance

---

[34] 45 SW History, CY 1993, Vol I, p 210; 45 SW History, CY 2007, Vol I, p 92.

[35] ESMC History, CY 1989, Vol I, pp 163, 166.

[36] ESMC History, CY 1989, Vol I, p 166.

connections and safety checks continued during the last week on the pad, and the vehicle was prepared for launch.[37]

Military Oversight

At this point one might ask, "*How* did the Air Force supervise DELTA II space operations at the Cape?" This is not an easy question to answer completely. Fortunately, launch vehicle and spacecraft training aids – maintained as Squadron Operating Instructions (SOIs) – were published shortly after military space operations were transferred to AFSPACECOM in October 1990. The instructions offered some basic terms of reference, and it is reasonable to assume that they were based on earlier 6555th Aerospace Test Group practices. As training aids the SOIs provide a brief, but detailed, look into the world of military space operations. The duties of 1 SLS officials (at least as they stood in the early 1990s) are presented below.[38]

Space Launch Squadron commanders had overall responsibility for launch operations in their squadrons, but they relied on a highly-trained and educated team of officers and non-commissioned officers to conduct day-to-day operations at the Cape. Space Launch Operations Controllers (SLOCs) acted as the Air Force's on-scene representatives during booster processing operations, and they ensured safety and security standards were maintained and proper procedures were followed to process launch vehicles and analyze any problems that surfaced during the course of operations. The SLOCs were authorized to stop operations any time procedures, safety guidelines or security standards were violated. In the event of an accident, the SLOCs took action to minimize injuries and equipment damage. They also preserved evidence of the mishap until relieved by proper authority.[39]

Systems Engineers (SEs) performed a variety of technical roles. They reviewed electronic and mechanical hardware modifications to hydraulic, pneumatic and propellant systems, structures, and ordnance. They monitored individual contractor actions as they occurred, and they reviewed and approved highly complicated procedures ahead of time. Their duties included participation in engineering "walkdowns" led by McDonnell Douglas launch operations managers to detect vehicle damage and improper hardware installation at the launch pads. During walkdowns, SEs immediately reported any discrepancies they detected, and they followed up with the appropriate company's engineering managers to resolve any problems that could not be corrected on the spot. In addition, an SE was selected to serve as a Vehicle Engineer (VE) for each launch vehicle. Vehicle Engineers monitored the status of all test procedures, vehicle problems, site concerns, and ground equipment issues. The senior VE briefed launch vehicle status at the Launch Readiness Review, and he tasked SEs to resolve technical problems. He also provided the engineering "Go/No-Go" recommendation to the Launch Controller (LC) on launch day.[40]

Countdown operations were among the most crucial activities in the entire launch preparation process. Squadron officials played important roles on the Air Force/McDonnell Douglas launch team responsible for those countdown sequences. The Squadron's portion of the team consisted of a Launch Director (LD), a Launch Controller (LC), a Launch Operations

---

[37] ESMC History, CY 1989, Vol I, pp 166, 167.

[38] Interview, M. Cleary with B/Gen. J. R. Morrell, 45 SW Commander, 9 March 1993; 1 SLS Squadron Operating Instruction (SOI) 55-1, 1 Oct 1991; 1 SLS SOI 55-2, 22 Mar 91; 1 SLS SOI 55-8, 4 Nov 1991.

[39] 1 SLS SOI 55-2, 22 Mar 1991, pp 1, 2.

[40] 1 SLS SOI 55-1, 1 Oct 1991, pp 3, 49, 50, 51; SLS SOI 55-8, 4 Nov 1991, p 43.

Manager (LOM), SLOCs, a Facility Operations Manager (FOM), a Booster Countdown Controller (BCC), SEs, a Facility Anomaly Chief (FAC), and an Anomaly Team Chief.[41]

The Launch Director reported directly to the Mission Director (MD) and commanded the launch crew through its pre-launch, launch, post-launch, and abort/launch scrub activities. The Launch Controller, the Anomaly Team Chief, and the Launch Weather Officer reported to the Launch Director. The Launch Director authorized continuation of the countdown following built-in holds that occurred during the countdown, and he also made the final launch vehicle Go/No-Go recommendation to the Mission Director before launch.[42]

The Launch Controller controlled operations at Space Launch Complex 17 during countdown preparations, terminal countdown, and the period the site was secured following countdown. The LC received status reports from McDonnell Douglas' Launch Conductor and other members of the Squadron's launch crew. Given the length of a typical DELTA II countdown, there were actually two LCs for each DELTA II mission: the first shift LC was on duty from the initial launch day crew brief through completion of initial checklist items. The second shift LC relieved the first shift LC and remained on duty until the launch complex was secured from countdown. On either shift, the LC was responsible for making sure the team was ready to move ahead through key milestones in the checklist. The LCs approved deviations in procedure, and they coordinated the team's reaction to local weather conditions and flight and ground system problems.[43]

The Launch Operations Manager provided contact between the LC in the blockhouse and SLOCs on the launch pad. The duty involved receiving and reporting pad and vehicle status to the LC before terminal countdown. The LOM verified the launch complex's readiness for launch operations. This included the complex's voice communications capability and its closed circuit television and camera coverage of the launch.[44]

The Facility Operations Manager was responsible for the launch complex's readiness (e.g., towers, fire control systems, electrical power, generators, fuel tanks, water systems and other support items). The Launch Base Services (LBS) [45] contractor assigned to the complex reported to the FOM, and the FOM controlled that contractor's operations on the pad, except for pad safety. The FOM verified the facility's readiness to support the Mobile Service Tower's rollback and the site's closeout before terminal countdown.[46]

---

[41] 1 SLS SOI 55-8, 4 Nov 1991, p 1.

[42] 1 SLS SOI 55-8, 4 Nov 1991, pp 2, 3.

[43] 1 SLS SOI 55-8, 4 Nov 1991, pp 9, 10.

[44] 1 SLS SOI 55-8, 4 Nov 1991, pp 16, 21.

[45] The Eastern Range was operated under a series of multi-year range contracts with Pan American World Services and its subcontractor, RCA, from 1954 through early October 1988. To improve competition for range services in 1988, the Air Force divided the range contract into the Range Technical Services (RTS) contract and the Launch Base Services (LBS) contract. Pan American (later known as Johnson Controls) was awarded the LBS contract in August 1988 and July 1992. The LBS contract was replaced with the Joint Base Operations and Support Contract (J-BOSC) in 1998. By 2007, the Air Force decided the J-BOSC was no longer the best way to handle services, so it began solicitations to replace the contract with a medley of 'breakout' contracts. The first of those contracts – for Infrastructure Operations and Maintenance Services – was awarded to InDyne, Inc., Reston, Virginia, on 16 June 2008.

[46] 1 SLS SOI 55-8, 4 Nov 1991, p 26.

For countdown operations, the Mission Vehicle Engineer (VE) was selected to be the launch team's Booster Countdown Controller (BCC) because he possessed the most comprehensive knowledge of the technical aspects of the launch vehicle's processing history (except for third stage and payload systems). The BCC controlled the Squadron's engineering team and ensured problems detected by the SEs on that team were brought to the attention of the Anomaly Team Chief and the Launch Controller. The SEs supervised the following operations during the countdown:[47]

    1. First stage propellant loading.

    2. Second stage propellant and pneumatic tank pressurization.

    3. Launch vehicle and ground support equipment electrical systems.
       (This area was assigned to two SEs.)

    4. Third stage electrical systems.

    5. Solid rocket motor, first and second stage pneumatic thruster pressure, and
       launch vehicle air-conditioning. (This area was assigned to two SEs.)

The Facility Anomaly Chief served as the Squadron's focal point for launch problems involving equipment and systems maintained by the LBS contractor. The FAC received countdown status from LBS contractor personnel and the Facility Operations Manager, and he reported technical concerns to the Anomaly Team Chief. The FAC was responsible for Space Launch Complex 17's electrical power, searchlights, pad deluge systems, and air-conditioning and ventilation systems.[48]

The Squadron Chief Engineer served as the Anomaly Team Chief (ATC) for each launch vehicle. The ATC was the single point of contact for vehicle problems. He evaluated the information he received from the FAC and reported problem solutions to the Launch Director. The ATC also reported on wind factors affecting the launch vehicle's trajectory, its structural load margins, and guidance control margins.[49]

Several squadrons currently assigned to the 45th Operations Group and 45th Launch Group worked very closely with the 1 SLS to ensure the success of DELTA II and DELTA III missions. Not all of them will be mentioned here, but the principal units include the 1st Range Operations Squadron (1 ROPS), the 45th Operations Support Squadron (45 OSS), and the 45th Launch Support Squadron (45 LCSS). Those units had antecedents under the 6555th Aerospace Test Group (6555 ASTG), the Eastern Space and Missile Center (ESMC), and the 45th Space Wing. Let's look at the antecedents first.

When the 6555 ASTG was placed under ESMC on 1 October 1979, the Group had three divisions to carry out its mission: 1) the Space Transportation System (STS) Division, 2) the Space Launch Vehicle Division, and 3) the Satellite Systems Division. Following the first Space Shuttle missions, the Air Force streamlined payload operations by consolidating the STS Division and the Satellite Systems Division. The two divisions became the Spacecraft Division on 1 November 1983. The next important shift in the 6555 ASTG's spacecraft operations occurred on 1 October 1990 when 175 out of 241 personnel were transferred 'on paper' to the 1 SLS, the

---

[47] 1 SLS SOI 55-8, 4 Nov 1991, pp 31, 37.

[48] 1 SLS SOI 55-8, 4 Nov 1991, p 49.

[49] 1 SLS SOI 55-8, 4 Nov 1991, p 43.

ATLAS II and TITAN IV Combined Test Forces, "Ops Resource Management," and "Payload Operations"[50] under ESMC. As noted earlier in this chapter, ESMC was transferred from Air Force Systems Command to Air Force Space Command on 1 October 1990, and the Center was groomed over the next 13 months to become the 45th Space Wing.[51]

As part of the inactivation of ESMC and its redesignation and activation as the 45th Space Wing on 12 November 1991, the Headquarters Eastern Test Range was redesignated the 45th Range Squadron (45 RANS). The 45th Operations Support Squadron (45 OSS) was reconstituted from the 45th Airdrome Squadron, an old inactive unit. When the 45 OSS was redesignated and activated, the Ops Resource Management Office became its launch operations support agency. The 45th Spacecraft Operations Squadron (45 SPOS) was newly activated, and it absorbed the resources of the Payload Operations Office. All three of those reconstituted, redesignated or newly activated squadrons were placed under the Headquarters, 45th Operations Group, the latest incarnation of the 45th Bombardment Group (Medium), which had been inactive since 8 December 1942.[52]

The 45 RANS planned and coordinated many of the actions – including allocation of resources – needed to carry out launch programs on the Eastern Range. Under the Headquarters ETR organization, the Directorate of Range Operations had accepted range users' requirements, monitored funding, and managed range support reimbursement programs. The ETR used Program Support Managers (PSMs) to monitor users' expenses and track the range contractor's performance. It also employed Range Control Officers (RCOs) to coordinate non-ESMC resources that supported the launch effort. Under the 45 RANS, PSM and RCO duties were merged and given to a new category of officials known as Range Operations Officers (ROOs). That new category of officer was assigned to the Squadron's ballistic operations, expendable space launch vehicle operations or Space Shuttle activities. The 45 RANS was inactivated on 1 December 2003, but its resources were transferred to the newly constituted and activated 1st Range Operations Squadron (1 ROPS) on the same day.[53]

Though it didn't remain active for very long, the 45th Spacecraft Operations Squadron (45 SPOS) played a pivotal role in military payload activities from mid-November 1991 until its inactivation on 13 May 1994. It also laid the foundation for supervision of payloads by other military agencies in later years. The 45 SPOS served as an executive agent for the Space and Missile Systems Center (SMC), and it processed Defense Department spacecraft for flights on ATLAS II, DELTA II, TITAN IV and Space Shuttle vehicles.[54] The Squadron also processed

---

[50] The Test Group's remaining personnel stayed where they were, but their numbers dwindled to approximately 25 military members and 11 civilians by the end of December 1991. The 6555 ASTG was deactivated on 1 July 1992.

[51] ESMC History, 1 Oct 1979 – 30 Sep 1981, Vol I, p 22; ESMC History, 1 Oct 1982 – 30 Sep 1984, Vol I pp 3, 4; ESMC/45 SW History, 1 Oct 1990 – 31 Dec 1991, Vol I, pp, 39, 40, 42, 43.

[52] AFSOC Special Order GD-030, 7 Nov 1991; ESMC/45 SW History, 1 Oct 1990 – 31 Dec 1991, Vol I, pp 1, 402, 403.

[53] ESMC/45 SW History, 1 Oct 1990 – 31 Dec 1991, Vol I, pp 20, 21, 24.

[54] On the contractor side of the 'payload' house, the McDonnell Douglas Space Systems Company (MDSSC) provided approximately 150 people under the Launch Operations Support Contract (LOSC) to support operations at the Spacecraft Processing and Integration Facility and to work at the Operations Support Center (OSC). Rockwell International Corporation provided 80 people to process NAVSTAR Global Positioning System satellites launched on DELTA II boosters. Elsewhere on the Cape, General Electric Astro Space employed 100 people to process Defense Satellite Communications System (DSCS) payloads launched on the ATLAS II, and Johnson Controls dedicated 28 people to facility maintenance and other ground support functions. Taken together, more than 400 military personnel,

payloads requested by other customers. On paper, the Squadron had separate flights for ATLAS II, DELTA II, TITAN IV and Shuttle payload operations, but only the DELTA Payload Operations Flight had dedicated officers and Non-Commissioned Officers (NCOs) – at least through 1992. The rest of the flights were supplied by a pool of officers and NCOs who were funneled into the separate flights as needed.[55]

The Air Force payload operations team was the linchpin for 45 SPOS spacecraft activities. The team was formed approximately one to two years before a spacecraft arrived at the Cape, and the team consisted of a Payload Operations Director (POD), a Field Program Manager and Deputy (FPM and DFPM), a Lead Non-Commissioned Officer (LNCO), Operations Controllers (OCs) and a Spacecraft Countdown Controller (SCC). The Payload Operations Director was responsible for all spacecraft launch base operations involving a particular booster class of payload. The POD managed and controlled spacecraft testing, fueling, pre-launch actions, and launch activities for the Mission Director. He also selected the Field Program Manager, who would keep the team updated on the spacecraft's status.[56]

The Field Program Manager exercised control over the contractor's field processing and launch activities. Those workloads stepped up dramatically about six to 12 months before a launch, and the FPM had to ensure that the launch base was ready to receive flight hardware at the proper time. Test procedures were reviewed and approved, facilities were configured, and the payload operations team was briefed. After the spacecraft arrived at the Cape (about six months before launch), the FPM chaired daily spacecraft scheduling meetings with the POD, the Deputy FPM, the Payload Support Contractor, the Launch Systems Integration Contractor, the Launch Vehicle Contractor, the SPO (System Program Office representative), the Air Force Launch Controller, the Aerospace Corporation's representative, and (often) a representative from Air Force Quality Assurance. Based on those meetings, the FPM approved the spacecraft schedule and made sure operations were performed according to that schedule. The Deputy FPM stepped in for the FPM if the latter was absent, and he attended to spacecraft security badging and test procedures.[57]

The Lead Non-Commissioned Officer was responsible for scheduling all range support activities for the spacecraft program. He received most of those requirements at the spacecraft scheduling meetings, but his experience and oversight were required to identify critical requirements that might have been missed in the documents presented at those meetings. The LNCO coordinated the spacecraft's arrival with the Operations Support Center and other support agencies approximately three days before the spacecraft arrived at the Cape. He was present during most major spacecraft operations. Through his contacts, he was able to get support for missed services – especially those identified as „show-stoppers.' The LNCO did not operate independently. He coordinated his requests with the Payload Support Contract Test Conductor. The work was scheduled through the Operations Support Center.[58]

---

civil servants and contractors supported spacecraft operations under the 45th Spacecraft Operations Squadron's taskings in the early 1990s.

[55] Interview, M. Cleary with Lt. Colonel Ivory J. Morris, 45 SPOS Commander, 19 Nov 1992; Slide Briefing, 45 SPOS, "Spacecraft, Pathway to the Future," o/a 15 Nov 1992.

[56] Interview, M. Cleary with Lt. Colonel Ivory J. Morris, 45 SPOS Commander, 19 Nov 1992; SOI 55-33, 45 SPOS, 16 Dec 1991; Self-Study Guide (SSG) A-15, 45 SPOS, 45 SPOS, "Review Operations Documents," 20 Aug 1992; SSG A-3, 45 SPOS, "Control Launch Base Operations," 14 Jul 1992, pp 1, 2, 3.

[57] SSG A-3, 45 SPOS, "Control Launch Base Operations," 14 Jul 1992, pp 1, 2, 3.

[58] *Ibid.*

The 45th Spacecraft Operations Squadron's Operations Controllers were responsible for controlling and supervising individual field operations (e.g., fueling, ordnance installation, solid motor buildup, and spacecraft lifts). Like the SLOCs in the launch squadrons, the spacecraft OCs had the authority to stop processes whenever safety or security standards were violated. The OCs ensured that „clean room' (contamination control) standards were maintained in the processing areas. In the event of a spacecraft anomaly or processing accident, the OC stopped the operation and alerted the FPM.[59]

This brings us to the final member of the payload operations team – the Spacecraft Countdown Controller. When qualified, the FPM served as the SCC on launch day. In any event, the SCC was responsible for the performance of pre-countdown and countdown procedures on major tests and launch operations. The SCC ensured spacecraft readiness, and he participated in trouble-shooting processes during the countdown. He reported to the Payload Operations Director and the Launch Controller, and he conferred with the Payload Support Contract Test Conductor and the SPO on issues that might warrant a launch hold. At predetermined points in the countdown, the SCC provided Go/No-Go recommendations including the final Go/No-Go recommendation before lift-off.[60]

On the morning of 13 May 1994, the 45th Spacecraft Operations Squadron was inactivated during a ceremony held in the Satellite Assembly Building (Building 49904) at Cape Canaveral. Following the inactivation, 45 SPOS resources were transferred to payload operations flights in the 1st, 3rd, and 5th Space Launch Squadrons. The transfer caused no significant changes in 45th Operations Group manning or grade structure. Military supervision of payloads continued under the launch squadrons.[61]

The next major organizational change affecting military oversight of DELTA II launch operations occurred on 1 December 2003. On that date, the 45th Launch Group (45 LCG) was constituted and assigned to Air Force Space Command with further assignment to the 45th Space Wing. Furthermore, the 1 SLS and 3 SLS were reassigned from the 45th Operations Group to the 45th Launch Group, and the 1st Range Operations Squadron (1 ROPS) was constituted and assigned to the 45th Operations Group. The 45th Maintenance Group was inactivated at Cape Canaveral Air Force Station, and the 45th Space Communications Squadron and 45th Range Management Squadron were reassigned from the (now defunct) 45th Maintenance Group to the 45th Operations Group. In addition, the 5 SLS – which had been inactivated since 29 June 1998 – was re-activated and assigned to the 45th Launch Group. As mentioned earlier, the 45 RANS stood down on 1 December 2003, but its resources were transferred to the 1 ROPS.[62]

Finally, the 3 SLS stood down on 30 June 2005, and the 45th Launch Support Squadron (45 LCSS) was activated to supervise spacecraft processing and provide "Group Level" mission support to the 45 LCG. Put simply, the 45 LCSS activation was designed to integrate the 45 LCG's payload activities under one roof. Consequently, the 1 SLS' six-member Spacecraft Flight

[59] SSG A-3, 45 SPOS, "Control Launch Base Operations," 14 Jul 1992, pp 3, 4, 5; Summary, Major K. W. Carlson, 45 SPOS, "Local Operations Requiring Controller Coverage," undated; SOI 55-21, 45 SPOS, "Payload Operations Control," 16 Dec 1991; SSG A-1, 45 SPOS, 7 Jul 1992, pp 11, 12; SOI 55-31, 45 SPOS, "Personnel Safety," 16 Dec 1991.

[60] SSG A-3, 45 SPOS, "Control Launch Base Operations," 14 Jul 1992, pp 3, 4; Slide Briefing, Major K. W. Carlson, 45 SPOS, "Introduction to Spacecraft Operations and Training," undated; SSG C-2, 45 SPOS, "Perform Countdown Activities," 2 Jul 1992, pp 3, 4.

[61] 45 SW History, CY 1994, Vol I, p 2; HQ AFSPC SO GD-016, 4 Apr 1994.

[62] HQ AFSPC SO GD-002, 24 Nov 2003; 45 SW History, CY 2003, Vol I, pp 2, 3.

was rolled into the new squadron. The 45 LCSS also provided additional manpower to ensure that training and programming services were available to support the Launch Group's mission and that mission-critical facilities would be available for space launch operations.[63]

Despite all the pulling and hauling of other squadrons from one group to another over the past 15 years, the 45th Operations Support Squadron continues to operate under the 45th Operations Group. It provides a wide variety of services, including the coordination and implementation of policies and procedures related to launch operations, resolution of operational issues with higher headquarters, coordination of launch operations, and instructor training for 45th Operations Group mission-ready personnel. In addition to those activities, the 45 OSS manages all airfield and air traffic control services for the 45th Space Wing. Last and far from least, the 45 OSS handles protocol issues, coordinates launch-critical briefings and conferences, and arranges tours and briefings for distinguished visitors.[64]

Officer, Enlisted and Civilian Figures for the 1 SLS and Related Units

Manpower figures for the 1 SLS, 45 SPOS, 45 OSS, 45 RANS, 1 ROPS and 45 LCSS – and for the 45th Operations Group and 45th Launch Group headquarters – varied over the years. Appendix B provides annual year-end strength figures on the number of officers, enlisted personnel and civilians assigned to the aforementioned units following the activation of the 1 SLS on 1 October 1990. Appendix A lists the names of unit commanders and their length of tenure, and Appendix C is a synopsis of all the DELTA II and DELTA III missions launched at Cape Canaveral from February 1989 through the end of March 2009. As of this writing, the final Air Force-sponsored DELTA II mission – NAVSTAR GPS IIR-21 – is expected to be completed sometime in August 2009.

Annual launch rates and year-end strengths show definite trends, but the reader is cautioned to consider the figures carefully before coming to any conclusions correlating the two. Some launch campaigns were inherently more difficult that others, regardless of how many or few people were assigned to each of the squadrons and groups at the time a particular vehicle was launched. Bad weather, safety concerns, launch vehicle problems and tardy payload deliveries could have a significant impact on the length of time it took to complete any of the missions. Sometimes several missions featuring the same type of payload or launch vehicle could be delayed by a single anomaly or hardware defect. Significant delays will be addressed in the next three chapters for military, civil and commercial DELTA II missions. All three DELTA III missions will be reviewed in Chapter IV.

---

[63] HQ AFSPC SO GD-012, 23 Jun 2005; 45 SW History, CY 2005, Vol I, pp 1, 2.

[64] Fact Sheet, "45th Operations Group," undated.

This emblem is shared by the 45th Operations Group and the 45th Launch Group. It was approved for the Air Force Eastern Test Range on 19 July 1967, and it has continued as the official emblem of the Eastern Space and Missile Center (ESMC) and the 45th Space Wing.

Significance: The shield is bordered in gold and divided into ultramarine blue and gold quadrants. Blue is used to symbolize the sky and space, and gold is used to symbolize the excellence required to conduct successful range operations. Dividing the shield horizontally, across its right half, is a line of "Ts" representing continuous testing of space vehicles. In the center of the shield, a large ultramarine and light blue globe represents Earth. A smaller globe, in the same colors, symbolizes the moon and other planets. Nine pimento red flight arrows indicate the normal equatorial departure routes for missiles and space vehicles on the Eastern Range. They also symbolize travel to other planets, as depicted by the smaller globe. Red was chosen for the flight arrows to indicate the stresses of launch and space flight and the heat of reentry into Earth's atmosphere. A string of white clouds across the center of the large globe represents abnormal conditions – weather and radiation – with which range personnel have to contend.

This emblem was approved for the 1st Space Launch Squadron on 1 July 1992.

Significance: Blue and yellow are the Air Force colors. Blue alludes to the sky, the primary theater of Air Force operations. Yellow refers to the sun and the excellence required of Air Force personnel. The black disc represents space and reflects the area of mission focus for the unit. The globe in space suggests the support the unit provides to United States Forces anywhere in the world via the Global Positioning System which is symbolized by the polestar. The flight symbol and contrail denote the dedication and technical expertise of the launch team personnel. The squadron's motto is "Ad Astra" (to the stars).

This emblem was approved for the Headquarters, Eastern Test Range (an ESMC unit) on 21 April 1987. The Headquarters, Eastern Test Range was redesignated the 45th Range Squadron (45 RANS) on 12 November 1991.

Significance: Symbolic of the Eastern Test Range, the emblem is in the shape of a shield bordered in gold. The central deltoid of ultramarine blue and white represents the military and civilian space programs supported by the Eastern Test Range. The globe of forest and leaf green represents the readiness of the Range to support worldwide operations on a 24-hour basis. The upper right is an Air Force yellow field symbolizing the radar operations (Callsign: Gold) of the Range for both launch operations and tracking of Earth-orbiting satellites, represented by the silver satellite passing through the field. The upper left is a silver field symbolizing the telemetry support (Callsign: Silver) provided by the Range during launch and on-orbit operations. The group of nine stars and planet represents the exploration of space: past, present and future which the Eastern Test Range supports. The surrounding border of gold represents the excellence of ETR personnel in performance of their mission.

This emblem was approved for the 45th Operations Support Squadron (45 OSS) on 3 January 1996.

<u>Significance</u>: Blue and yellow are the Air Force colors. Blue alludes to the sky, the primary theater of Air Force operations. Yellow refers to the sun and the excellence required of Air Force personnel. The green planet alludes to the healthy Earth and the fact that the unit is environmentally conscious. The gridlines on the planet denote global coverage and continuing operations around the world. The winglets on the planet depict Airfield Operations. The four stars represent the four flights within the squadron. The rocket and contrail symbolize launch support and exploration of space

This emblem was approved for the 1st Range Operations Squadron (1 ROPS) on 14 May 2004.

<u>Significance</u>: Ultramarine blue and Air Force yellow are the Air Force colors. Blue alludes to the sky, the primary theater of Air Force operations. Yellow refers to the sun and the excellence required of Air Force personnel. The black field denotes space, an important frontier for mankind. The globe and green segmented field are representative of the unit's readiness to use its range instrumentation to achieve all duties required. The flight symbol represents all past, present and future space launches.

This emblem was approved for the 45th Launch Support Squadron (45 LCSS) on 28 December 2005.

<u>Significance:</u> Ultramarine blue and Air Force yellow are the Air Force colors. Blue alludes to the sky, the primary theater of Air Force operations. Yellow refers to the sun and the excellence required of Air Force personnel. The black background represents space. The three deltas signify the multiple launch vehicles the squadron supports and the three Air Force core values: Integrity First, Service Before Self, and Excellence in All We Do. The Earth and orbiting satellite describes the global nature of the unit's mission.

CHAPTER II

MILITARY MISSIONS

As of this writing, approximately half of all the DELTA IIs launched from Cape Canaveral have supported military missions (54/107). The vast majority of those flights – 48 – featured Global Positioning System (GPS) payloads. Three military DELTA II missions were dedicated to U.S. military technology experiments. Two DELTA II missions featured North Atlantic Treaty Organization (NATO) satellites. One DELTA II carried Britain's SKYNET 4D communications spacecraft into orbit.

Fifty-three of the flights were successful. Only one GPS IIR satellite mission failed, just seconds after the vehicle was launched from Pad 17A on 17 January 1997. There was considerable property damage in the immediate area as a result of the mishap, but there were no significant injuries. Remarkably, the next DELTA II mission was launched from the same site just four months later. Three more DELTA II missions were launched from Complex 17 before the end of 1997.

Successful or not, every DELTA II launch campaign demanded the launch contractors' best efforts. Engineers and technicians measured their progress by the safe and successful completion of commonly accepted milestones under the watchful eye of company supervisors, managers and government officials. Major milestones in the process included: 1) erecting the first stage and interstage at the launch site, 2) attaching the solid rocket motors, 3) erecting the second stage, 4) mating the payload to the launch vehicle, 5) installing the payload fairing, 6) completing combined systems tests and simulations, and 7) fueling the vehicle in preparation for the actual launch.

Based on our records, a „blow-by-blow' accounting of those milestones could be presented below, but the repetitive nature of that story would soon lose its luster. Instead this summary will merely highlight any significant delays that occurred in the course of the various missions. The first GPS mission was summarized in Chapter I, so follow-on GPS missions will be considered in this chapter. The three technology experiments will be covered here as well, and we will conclude this chapter with the DELTA II missions sponsored by NATO and the United Kingdom.

Early NAVSTAR II GPS Missions

Following the first NAVSTAR II GPS launch on 14 February 1989, 16 highly successful NAVSTAR II missions were launched from Complex 17 between 10 June 1989 and the end of 1992.[1] Though all 16 missions went well, countdowns did not always lead to launches. The NAVSTAR II-2 mission was one of many to encounter delays at the launch site. It was launched from Pad 17A at 2230:00.734Z on 10 June 1989, but only after range officials were forced to scrub five earlier launch attempts.[2] The NAVSTAR II-3 mission encountered similar problems. It was scheduled to lift off on 11 August 1989, but it was delayed 24 hours after Pad 17A suffered a lightning strike. Its first countdown got underway on 12 August, but officials had to scrub the

---

[1] Model 6925 DELTA II launch vehicles were used for the GPS II-1 through GPS II-9 orbital missions. Later missions required more powerful Model 7925 vehicles to lift heavier Block IIA and IIR satellites.

[2] Countdowns for the NAVSTAR II-2 launch were attempted on 20 May, 21 May, 23 May, 24 May and 9 June 1989 before the mission was launched on 10 June. A liquid oxygen (LOX) valve problem prompted the launch scrub on 24 May, and the other launch attempts were scrubbed due to weather constraints.

launch due to a weather constraint violation. The next countdown got underway six days later, and the DELTA II lifted off Pad 17A at 0557:58.876Z on 18 August 1989.[3]

On some occasions, only one countdown was needed to launch a NAVSTAR II mission. Other countdowns might be recycled and completed successfully within a few days. The NAVSTAR II-4 mission, for example, was launched at the end of its first countdown at 0931:00.735Z on 21 October 1989. Officials scrubbed the NAVSTAR II-5 mission on 10 December 1989 due to a helium problem in the vehicle's second stage, but the vehicle lifted off Pad 17A the next day at 1810:00.793Z. NAVSTAR II-6 was launched on its first try at 2255:00.531Z on 24 January 1990. Officials scrubbed the NAVSTAR II-7 launch on 21 March 1990 due to unacceptable upper level wind conditions, but the vehicle was launched successfully from Pad 17A at 0245:00.751Z on 26 March 1990. The NAVSTAR II-8 payload was launched after an uneventful countdown at 0538:59.838Z on 2 August 1990. A DELTA II boosted the NAVSTAR II-9 payload into orbit from Pad 17A at 2156:00.023Z on 1 October 1990.[4]

The Model 7925 DELTA II made its debut flight from Pad 17A at 2139:01.144Z on 26 November 1990. Apart from a seven-minute extension to evaluate wind data, the countdown was uneventful. The new vehicle placed the first GPS IIA spacecraft (NAVSTAR II-10) in the prescribed orbit. NAVSTAR II activities paused for the next several months while McDonnell Douglas concentrated its efforts on four separate launch campaigns on Pad 17B,[5] but the next NAVSTAR mission was prepared for launch shortly thereafter. Officials scrubbed the NAVSTAR II-11 mission on 3 July 1991 due to the loss of a satellite communication link with the range tracking station on Ascension Island. The countdown was picked up again the next day, and the GEM-equipped DELTA II boosted its NAVSTAR II-11 payload and the LOSAT-X payload [6] off Pad 17A at 0232:00.057Z on 4 July 1991. Both payloads were released successfully. The first launch attempt for the NAVSTAR II-12 mission was aborted at 2322:00Z on 18 February 1992 due to heavy cloud cover, and the second countdown was scrubbed at 2310:00Z on 19 February for weather problems. The third and final attempt for that mission on 23 February was overshadowed by bad weather and punctuated with a brief computer synchronization problem, but the countdown was otherwise uneventful. The weather improved, and the DELTA II lifted off Pad 17B successfully at 2229:00.004Z on 23 February 1992.[7]

---

[3] ESMC History, FY 1989, Vol I, pp 364, 365, 366.

[4] ESMC History, FY 1990, Vol I, pp 311, 312, 313, 314; ESMC/45 SW History, 1 October 1990 - 31 December 1991, Vol I, p 337.

[5] McDonnell Douglas launched the NATO IV-A spacecraft from Pad 17B on 8 January 1991. See *Foreign Military Missions* later in this chapter. Three commercial spacecraft – INMARSAT-2 F-2, ASC-2, and AURORA II – were launched from Pad 17B on 8 March, 13 April, and 29 May 1991 respectively.

[6] The LOSAT-X should have been launched along with two other Strategic Defense Initiative payloads (e.g., LOSAT L and LOSAT R) on a DELTA II launch vehicle on 14 February 1990. Unfortunately, the "X" payload was not ready in time for that launch, so it was re-manifested as part of the NAVSTAR II-11 flight. The LOSAT-X was mounted sideways on the DELTA II's second stage, and it was jettisoned approximately one hour into the flight – about 40 minutes after second stage/third stage separation. A ground station in Boulder, Colorado, provided primary control for the LOSAT-X on its classified mission. The Consolidated Space Test Center at Sunnyvale, California, provided a backup control capability.

[7] ESMC/45 SW History, 1 October 1990 - 31 December 1991, Vol I, pp 337, 338, 339; 45 SW History, CY 1992, Vol I, pp 230, 232; Range Pretest Briefing, CSR, "DELTA II LOSAT," 26 Jan 1990, p 2.

The next NAVSTAR II mission also required three countdowns [8] before its Model 7925 vehicle lifted off Pad 17B on 10 April 1992. The NAVSTAR II-14 mission was launched from Pad 17B on the first launch attempt at 0920:00.574Z on 7 July 1992. Similarly, the NAVSTAR II-15 mission was launched from Pad 17A on its first launch attempt at 0857:00.267Z on 9 September 1992. Officials scrubbed the NAVSTAR II-16 countdown on 7 November 1992 at 0105Z due to a vehicle misfire,[9] and they scrubbed the mission's next countdown on 20 November at 0005Z due to weather constraints. In contrast to those first two attempts, the countdown for the NAVSTAR II-16 launch on 22 November went smoothly. There were no unscheduled holds, and the vehicle lifted off Pad 17A without incident at 2354:00.001Z on 22 November 1992. Officials scrubbed the final NAVSTAR II mission of 1992 on 16 December due to a vehicle anomaly that occurred during the final liquid oxygen filling procedure. The second countdown on 18 December 1992 was successful, and the DELTA II lifted the NAVSTAR II-17 payload off Pad 17B at 2216:00.10Z on that date. The flight was the 17th in an unbroken series of successful missions for the NAVSTAR II GPS program.[10]

The DELTA II/NAVSTAR II launch program was one the Air Force's greatest success stories at the Cape, so it is only fitting that we should note its impact on the NAVSTAR GPS constellation in the late 1980s and early 1990s. Of the 10 Block I NAVSTAR spacecraft orbited before November 1985, seven were still in operation when the first NAVSTAR II mission was launched in February 1989.[11] In 1988, the Air Force and the Joint Requirements Oversight Committee of the Joint Chiefs of Staff supported a return to the original plan of a 24-satellite GPS constellation. On 27 February 1989, the Air Force directed Air Force Systems Command to establish and maintain a GPS constellation of 21 primary satellites, plus active orbiting spares (i.e., a 24-satellite constellation). The last two NAVSTAR II satellites were removed from the Space Shuttle's payload manifest in the spring of 1989, so the entire weight of the expanded constellation fell on the shoulders of the DELTA II launch program.[12]

The GPS Program Office hoped to have five NAVSTAR II satellites in orbit by the end of September 1989, but only three of those spacecraft had been launched by that time. Since 12 Block II/IIA satellites would be needed to give the GPS constellation its first worldwide two-dimensional navigation capability, planners estimated that capability could not be achieved before the spring of 1991.[13] In point of fact, five more NAVSTAR II GPS spacecraft were launched from Pad 17A by early August 1990, and Iraq's invasion of Kuwait in that month provided additional incentive for McDonnell Douglas and the Air Force to launch more GPS satellites.[14]

---

[8] The first two countdowns were scrubbed on 4 and 5 April 1992 due to upper level winds. Weather was also a concern during the night of 9 April, but wind conditions improved. The third and final countdown was uneventful, and the vehicle lifted off the pad at 0319:59.988Z on 10 April 1992.

[9] The DELTA II vehicle was placed in a safe condition after the misfire, and it took two hours to de-tank the kerosene and liquid oxygen propellants from the vehicle's first stage.

[10] 45 SW History, CY 1992, Vol I, pp 232, 234, 236, 237, 238

[11] One of the seven – NAVSTAR 4 – was rated marginal due to subsystem failures, and it needed extensive support from ground controllers to remain in operation during 1989. It was finally removed from the operational GPS constellation in August 1989.

[12] 45 SW History Office, "The Cape: Military Space Operations, 1971 – 1992," p 172.

[13] The GPS constellation consisted of six Block I and three Block II satellites before five more Block II spacecraft were added to the constellation in FY 1990.

[14] 45 SW History Office, "The Cape: Military Space Operations, 1971 – 1992," pp 172, 173.

NAVSTAR II-9 (the last of the Block II satellites) lifted off Pad 17A on 1 October 1990, and it was placed in orbit over the Middle East. The spacecraft's on-orbit testing program was completed in record time, and NAVSTAR II-9 was turned over to Air Force Space Command on 24 October 1990. NAVSTAR II-10 was launched successfully on 26 November 1990. With II-10 in operation, the GPS network provided two-dimensional coordinates with an average accuracy of roughly five meters during Operation DESERT STORM. The NAVSTAR system's three-dimensional accuracy averaged about 10 meters during the war. The GPS Program Office hoped to have five Block IIA NAVSTAR spacecraft on-orbit by October 1991, but component problems associated with the new design caused lengthy delays. Only two Block IIA missions were launched by October 1991, but six more Block IIA launches were completed by the end of 1992. The constellation was well on its way to full operational status by the beginning of 1993.[15]

Completing the Initial NAVSTAR II Constellation

Between 3 February 1993 and mid-March 1994, McDonnell Douglas launched seven GPS missions (NAVSTAR II-18 through II-24) to deliver the initial NAVSTAR II constellation on-orbit. Four more Block IIA missions (GPS II-25 through GPS II-28) were launched in 1996 and 1997 to replenish the constellation. The first two satellites in the new GPS IIR spacecraft series were launched in 1997. All the missions except GPS II-25 were launched from Pad 17A, and all of the spacecraft except GPS IIR-1 were boosted into 10,898-nautical-mile-high transfer orbits to become part of the existing GPS navigational satellite network.

The first of those missions – NAVSTAR II-18 – was launched on 3 February 1993. Coincidentally, 45th Space Wing's official histories began featuring processing milestones for each launch vehicle's build-up around that time. Consequently, a milestone summary for the NAVSTAR II-18 mission is available, and an edited version of it has been provided below to show how milestones were completed during a somewhat typical DELTA II launch campaign in the 1990s.

> As part of the NAVSTAR II-18 effort, contractors delivered the DELTA II's first stage to Cape Canaveral on 29 September 1992. Technicians completed the first stage's receiving inspection on 3 October 1992. The payload fairing and second stage were on station by that time, and officials completed the receiving inspections for the payload fairing and second stage by the end of September 1992. Technicians and engineers accomplished interstage processing on 21 and 22 October 1992. Workers processed the payload fairing between 1 and 12 December 1992. Engineers and technicians erected the DELTA II's first stage and interstage on Pad 17A on 21 December, and the second stage and payload fairing were in place by 24 December 1992.[16]

> The spacecraft arrived at the Cape on 15 April 1992, more than four months ahead of its launch vehicle. Technicians completed the spacecraft receiving inspection on 20 April 1992. Then the satellite was placed in storage at the NAVSTAR Processing Facility near the Skid Strip at Cape Canaveral. Engineers began processing the payload for the mission on 22 October 1992, but the work was halted on 20 November due to a flaw in the satellite's Burst Detector Processor (BDP). The BDP was shipped to Sandia National Labs in New Mexico for repair, and spacecraft workers had to put in many 12-hour workdays to get the payload back on schedule once the BDP was returned about three weeks later.[17]

---

[15] 45 SW History Office, "The Cape: Military Space Operations, 1971 – 1992," p 173.

[16] 45 SW History, CY 1993, Vol I, pp 210, 211.

[17] 45 SW History, CY 1993, Vol I, p 211.

Officials held an Interim Readiness Review (IRR) on 20 January 1993 to confirm the payload and its PAM-D upper stage were ready to be mated. Technicians and engineers completed the mating operation on 20 January 1993, and they finished up their closeouts and packaging procedures two days later. The spacecraft was transported to Pad 17A early on the morning of 25 January, and technicians and engineers mated the payload to the DELTA II launch vehicle shortly thereafter. Technicians installed the payload fairing on 29 January, and officials completed the Launch Readiness Review on 1 February 1993. The countdown got underway the next day.[18]

High surface winds delayed the Mobile Service Tower rollback for approximately one hour, but there were no unscheduled holds during the countdown. The DELTA II lifted off the launch pad at 0255:00.174Z on 3 February 1993. The NAVSTAR GPS II-18 mission was a success.[19]

The NAVSTAR II-19 launch was the next mission in line. Some delays were encountered during II-19's preparation for flight, though none of the hold-ups were serious. The payload featured the Small Expendable Deployer System (SEDS) in addition to a standard Block IIA spacecraft.[20] The first countdown got underway on 18 March 1993, but officials scrubbed the launch when high winds prevented removal of the Mobile Service Tower (MST). They also had to scrub the second launch attempt on 19 March due to air-conditioning problems on the launch pad. The launch was rescheduled, but it was delayed until 28 March to give authorities time to review data concerning the abnormal operation of a Rocketdyne engine during the ATLAS AC-74 flight, which was launched from the Cape on 25 March 1993.[21] Officials scrubbed the DELTA II's third launch attempt on 28 March due to high upper level winds, but the countdown on 29/30 March finally did the trick. There were no unscheduled holds, and the DELTA II lifted off the launch pad at 0308:59.824Z on 30 March 1993.[22]

The contractor delivered the first stage of the NAVSTAR II-20 launch vehicle to the Cape on 8 December 1992, and the second stage arrived two weeks later. The satellite arrived on 13 January 1993. Launch delays and high winds held up the mission for seven days in April, but engineers mated the GPS satellite to the DELTA II launch vehicle on the morning of 4 May 1993. Officials started the countdown for the launch on 12 May 1993. There were no unscheduled holds during the count, and the DELTA II lifted off the launch pad at 0006:59.751Z on 13 May 1993.[23]

NASA's 146-pound Plasma Motor Generator (PMG) was included as a secondary payload[24] on the next NAVSTAR mission in late June 1993. The contractor delivered the DELTA

---

[18] 45 SW History, CY 1993, Vol I, p 211.

[19] 45 SW History, CY 1993, Vol I, pp 211, 212.

[20] The SEDS rode along as a small secondary payload, and it was released from the DELTA II's second stage guidance section as a modest experiment. It was a 30-pound "spinning reel" designed to deploy a 50-pound instrumented payload on the end of a 12.4-mile-long tether downward (i.e., toward Earth) to investigate the dynamics of low-tension deployment on a non-conducting tether. The instrumented payload contained sensors to measure temperature, attitude, force and acceleration. Data was stored in a small onboard computer for transmission on a single 2260.0 MHz S-Band telemetry channel.

[21] Since Rocketdyne manufactured liquid engines used on both the DELTA II and the Atlas I, McDonnell Douglas and the Air Force wanted an additional day of review data to confirm there was no connection between the recent Atlas I flight anomaly and any engines used on DELTA II missions.

[22] 45 SW History, CY 1993, Vol I, pp 212, 213.

[23] 45 SW History, CY 1993, Vol I, p 214.

[24] The PMG consisted of a "near end package" and a "far end package" similar to the SEDS-1 launched on the NAVSTAR II-19 mission in March 1993.The PMG was attached to a 1600-foot-long conductive tether to see if: 1) the tether could be charged magnetically as the PMG flew through the ionosphere, and 2) the magnetic charge could be used to drive the tether forward or backward. The ultimate purpose of the PMG experiment was to test the feasibility of using

II's second stage on 4 December 1992, and the first stage arrived on 27 January 1993. The spacecraft arrived at the Cape on 13 January 1993. Launch vehicle and spacecraft processing operations went smoothly, and technicians and engineers mated the GPS payload to its DELTA II launch vehicle on 15 June 1993.[25]

The countdown for NAVSTAR II-21 got underway on the morning of 26 June 1993, and it was memorable, to say the least. The launch was put on hold for thunderstorms and lightning at 1234Z, and the Mission Director decided to delay the launch until the second launch window (e.g., 1309Z) while the weather cleared. In the meantime, four intruder aircraft were detected in the launch danger zone around 1250Z, and sport fishing boats from a local tournament wandered into the launch danger area about 10 minutes later. Officials extended the hold at T minus 4 minutes to get both types of unauthorized craft out of the area, and they coordinated a new lift-off time of 1327Z. The countdown resumed at 1323Z, and the launch vehicle lifted off the launch pad at 1326:59.740Z on 26 June 1993. The DELTA II lagged somewhat in flight, but the payload was inserted in the proper transfer orbit.[26]

The NAVSTAR II-22 mission was anticipated with great interest, and contractors delivered the spacecraft and the DELTA II's second stage on 13 October 1992 and 5 November 1992 respectively. Unfortunately, the first stage did not arrive until 23 March 1993, and the spacecraft was shipped without its Burst Detector Processor, Reaction Wheel Assembly, and three SAFT batteries. (The vendor had to forward those items separately at a later date.) Despite those annoying delays, technicians and engineers managed to erect the DELTA II on the launching pad by 2 August 1993. In the meantime, the spacecraft remained in storage at the NAVSTAR Processing Facility (NPF) until 20 April 1993. Officials completed Master Control Station and Mission Control Center compatibility tests on the satellite on 4 June and 11 June 1993, and the spacecraft was fueled at the Propellant Servicing Facility by 6 August 1993. The payload was mated to the DELTA II on the morning of 20 August 1993.[27]

The launch was scheduled for 2 September 1993, but processing was wrapped up a few days early. Capitalizing on the saved time, officials decided to move up the launch to 30 August 1993. That action allowed an early launch opportunity for the next ATLAS II/CENTAUR mission elsewhere on the Cape, and it avoided a potential threat from Hurricane EMILY. There were no unscheduled holds during the countdown, and the DELTA II lifted off Pad 17A at 1238:00.092Z on 30 August 1993.[28]

The contractor delivered the second stage and first stage for the NAVSTAR II-23 mission on 13 July 1993 and 9 August 1993 respectively. The spacecraft selected for the mission had arrived back in April 1992 – presumably to support of the NAVSTAR II-21 mission – but, as luck would have it, a hardware problem surfaced during initial spacecraft processing. Consequently, officials transferred the spacecraft to the NAVSTAR II-23 manifest and placed the spacecraft in

---

this method of magnetic conduction to boost low-Earth orbiting satellites into slightly higher orbits. The Lewis Research Center in Cleveland, Ohio, was responsible for the PMG's plasma diagnostics package. The Marshall Space Flight Center was responsible for the electronics interface between the payload and the DELTA II's second stage power and telemetry systems. The Johnson Space Center served as lead center for the experiment.

[25] 45 SW History, CY 1993, Vol I, p 215.

[26] 45 SW History, CY 1993, Vol I, p 216.

[27] 45 SW History, CY 1993, Vol I, p 217.

[28] 45 SW History, CY 1993, Vol I, pp 217, 218.

storage pending processing in late November 1992.[29] Telemetry measurements later revealed that the satellite's solar array drive was in the wrong position. The problem took months to correct, but technicians and engineers eventually mated the spacecraft to the launch vehicle on 16 October 1993. Technicians and engineers had finished assembling the launch vehicle on the pad in late September 1993, so the team was eager to launch.[30]

Only one countdown was required to get the NAVSTAR II-23 payload into orbit. Low cloud cover caused a 21-minute unscheduled hold during the countdown, but the cloud ceiling lifted. The DELTA II's lifted off the pad at 1704:00.052Z on 26 October 1993.[31]

Technicians and engineers erected the DELTA II's first and second stages for the next NAVSTAR GPS II flight on 24 January 1994 and 10 February 1994 respectively. Reminiscent of the DELTA II GPS mission back in March 1993, this mission (II-24) featured a Small Expendable Deployer System (SEDS) in addition to a standard 4,150-pound Block IIA spacecraft. A laser retro-reflector was added to the forward bulkhead of the spacecraft to provide more accurate laser ranging measurements,[32] and the SEDS-2 was deployed after the Block IIA spacecraft separated from the launch vehicle.[33]

The payload contractor delivered the II-24 satellite to the Skid Strip at Cape Canaveral on 21 April 1993. Rockwell and McDonnell Douglas employees mated the Block IIA spacecraft to its PAM-D upper stage on 22 February 1994. The payload was installed in its "handling can" and transported to the pad, where engineers and technicians installed it atop the DELTA II on 1 March 1994. (The SEDS-2 was mated to the DELTA II's second stage on the evening of the 1st.) The payload fairing was installed on 5 March, and officials prepared for the launch.[34]

The countdown began with preparations for the Mobile Service Tower's rollback at 1130 Eastern Standard Time (EST) on 9 March 1994. The countdown proceeded smoothly to the 10-minute built-in hold at T minus 4 minutes. Officials extended the T minus 4-minute hold eight additional minutes to discuss the second stage's nitrogen bottle temperature limit, but the countdown continued after that issue was resolved. The DELTA II vehicle lifted off the launch pad at 0340:00.931Z on 10 March 1994. The spacecraft separated from the launch vehicle approximately half an hour after lift-off. The Air Force Satellite Control Network (AFSCN) assumed control of the satellite shortly after separation. Following on-orbit checkout, officials declared the NAVSTAR II-24 operational in April 1994.[35]

The next three GPS missions (II-25 through II-27) were launched in 1996 to ensure undiminished three-dimensional precise navigation support from the NAVSTAR II constellation. Apart from a three-day processing delay caused by the unrealized threat of Hurricane BERTHA in early July 1996, the first two launch campaigns advanced in routine fashion. NAVSTAR II-26 was the first military mission to employ the Redundant Inertial Flight Control Guidance Assembly

---

[29] A new spacecraft was substituted for the NAVSTAR II-21 mission. It arrived at the Cape on 13 January 1993.

[30] 45 SW History, CY 1993, Vol I, p 218.

[31] 45 SW History, CY 1993, Vol I, p 219.

[32] The retro-reflector was introduced on the NAVSTAR GPS II-22 mission, which was launched on 30 August 1993.

[33] 45 SW History, CY 1994, Vol I, pp 106, 107.

[34] 45 SW History, CY 1994, Vol I, p 107

[35] 45 SW History, CY 1994, Vol I, pp 107, 109.

(RIFCA).[36] NAVSTAR II-25 and II-26 were launched successfully at 0021:21.011Z on 28 March 1996 and 0050:00.073Z on 16 July 1996 respectively.[37]

The NAVSTAR II-27 mission also went well, though a lightning strike at the Range Operations Control Center (ROCC)[38] at 2039Z on 11 September 1996 switched that facility from critical power to industrial power. Another lightning strike dropped the ROCC's power down to UPS (Uninterruptible Power Source) emergency backup power. The incident was short-lived, and technicians restored critical power to the facility at 2258Z. There were no unscheduled holds during the countdown, and the DELTA II lifted off the launch pad at 0849:00.161Z on 12 September 1996. The 1 SLS dedicated the launch to America's prisoners of war and soldiers missing in action (POW/MIA). Many former POWs and their spouses who were visiting the area for an American ex-POW Association conference witnessed the successful launch.[39]

The NAVSTAR GPS IIR -1 Mission, 17 January 1997

The object of this mission was to place the first Block IIR replenishment satellite in a transfer orbit bound for the NAVSTAR II constellation 10,898 nautical miles above Earth. Lockheed Martin Missiles and Space (LMMS) provided the 4,480-pound replenishment spacecraft, which had an operational life expectancy of 10 years. The IIR series provided increased navigation accuracy, and the new spacecraft were better at station-keeping, thus reducing the necessity of ground control corrections. The new satellites were fully compatible with GPS satellites already on-orbit. In all, LMMS planned to deliver 21 new IIR-series spacecraft to replenish the aging GPS constellation between 1997 and 2006.[40]

Engineers and technicians erected the DELTA II first stage and interstage on Pad 17A on 5 December 1996. They erected the second stage on the following day. Problems with the Solid Motor Tunnel Covers and the Solid Rocket Hoist delayed the erection of the vehicle's GEMs until the 26th, but technicians erected all nine GEMs between 26 and 29 December 1996. On 31 December, Range Safety granted approval for a launch attempt on 16 January 1997. Technicians and engineers mated the spacecraft to the vehicle on 2 January, and the payload fairing was installed on 8 January 1997. Following second stage propellant loading operations on 13 January, technicians installed "Class A" ordnance in the Flight Termination System. Officials made final preparations for the launch.[41]

There were no unplanned holds during the countdown on the 16th, but officials scrubbed the launch for upper level winds. They recycled the count 24 hours, and they prepared for the launch once again. The vehicle lifted off the launch pad without incident at 1628:00.657Z on 17 January 1997, but the DELTA II was headed for destruction. Approximately 7.2 seconds after ignition, GEM #2 developed a 71-inch-long split in its casing. The split grew to 254 inches before

[36] McDonnell Douglas employed the RIFCA for the first time on the DELTA II Rossi X-Ray Timing Explorer (RXTE) flight on 30 December 1995. The RIFCA employed redundant laser gyros and accelerometers to sense the DELTA II's velocity, angular position and direction. In conjunction with a master telemetry unit and a power and control box, the RIFCA replaced a 1970s vintage avionics suite used on earlier DELTA launches.

[37] 45 SW History, CY 1996, Vol I, pp 92, 93, 94, 95.

[38] The ROCC was renamed the Morrell Operations Center (MOC) in honor of Major General Jimmey R. Morrell, former 1st Space Division, ESMC, and (later) 45th Space Wing Commander, on 2 November 2007.

[39] 45 SW History, CY 1996, Vol I, pp 95, 96; 45 SW History, CY 2007, Vol I, p 31.

[40] 45 SW History, CY 1997, Vol I, p 93.

[41] 45 SW History, CY 1997, Vol I, pp 93, 94.

the motor failed catastrophically about five seconds later (e.g., T plus 12.6 seconds). The casing failure prompted the first stage automatic destruct system, which destroyed the vehicle's first stage. The second stage, third stage and payload remained largely intact.[42]

Observing those developments, Mission Flight Control Officers (MFCOs) sent command destruct functions to control the disintegration of the vehicle at approximately T plus 22.3 seconds. Their actions destroyed the second and third stages, which, in turn, released the payload fairing and payload. Unfortunately, the payload and fairing exploded on impact with the ground.[43]

According to McDonnell Douglas' subsequent analysis of the mishap, there were no telemetry or visual indications of the explosion prior to the actual event. Weather was not a factor, and the cause of the explosion was unknown. Regarding events immediately after the mishap, the vehicle was about 1,590 feet above the ground and 100 feet downrange when the explosion occurred. The detonation of GEM #2 destroyed a GEM next to it, and the automatic destruct system took out the remaining GEMs and the first stage. Between 2,000 and 2,500 "firebrands" were released by the exploding GEMs and another 2,100 fragments were released by the disintegrated vehicle.[44] Many of those firebrands left craters, and four fragments (e.g., the payload/PAM-D motor and three large GEM pieces) caused secondary explosions estimated at between 1,250 and 2,000 pounds of TNT. Workers found small fragments and unburned pieces of solid propellant as far away as the USAF Space & Missile Museum and the north end of Complexes 31 and 32, but all debris fell well within the Flight Hazard Area, up to 6,500 feet from the launch pad. The toxic cloud from the explosion drifted quickly out to sea.[45]

No one was killed or injured as a result of the accident, but 26 vehicles, including a tractor trailer and a golf cart, were totally destroyed. Forty-six other vehicles were damaged. Four modular trailers were destroyed, and seven others received some degree of damage. The launch pad was not seriously damaged, and the Mobile Service Tower and Umbilical Tower sustained no more damage than they experienced during a normal DELTA II launch. Private property damage (including the leased trailers and the vehicles mentioned above) came to approximately $429,000.[46]

There were 73 people in the blockhouse when the accident occurred. A large piece from one of the GEMs landed on the northeast corner of the blockhouse. The explosion caused damage to the protective berm, but it did not penetrate the blockhouse.[47] The occupants were shielded from the fire caused by the explosion, though an appreciable amount of smoke filtered in via a

---

[42] 45 SW History, CY 1997, Vol I, p 94.

[43] *Ibid.*

[44] The GEM firebrands averaged about 0.5 cubic feet in size each, and they caused small fires throughout the Flight Hazard Area. The unpressurized GEMs, however, broke into several major pieces and caused secondary explosions and sizeable craters. The PAM-D third stage motor also produced a major secondary explosion equal to about 2,000 pounds of TNT. It left a crater 27 feet in diameter.

[45] Excerpt, Mr. Stu Backus, MDAC, "Mechanical Ordnance Event Scenario," undated; Message, 45 SW/SEG, "Class A, Space, Preliminary Report, 97/01/17, DBEH 001A, DELTA Launch Vehicle with GPS 2R Exploded In Flight," 172215Z Jan 97; Study, McDonnell Douglas, "Debris Study for DELTA II, GPS IIR-1, OPNR A4949 from Cape Canaveral AS, Pad 17A," ca 1 Mar 1997, pp 1, 4, 5, 13, 21, A-1

[46] Summary, Range Safety, "DELTA II-241 Class A Mishap, 17 Jan 1997, Estimated Cost of Private Property Damage," undated.

[47] Photos taken of the area after the accident looked worse than the actual damage caused by the GEM because a lot of water was used to put the fire out; the water washed away the berm in the northeast corner of the blockhouse's perimeter.

cableway. As one observer noted, the inside the blockhouse took on the appearance of a "smoky bar." All occupants donned breathing apparatus, and firefighters and emergency response teams escorted them out of the area. There were no injuries due to smoke inhalation.[48]

According to a debris study completed during the investigation of the accident, "buildup of the berm around the Blockhouse saved the people within from any serious damage by the piece of propellant that impacted the cableway on the eastern edge of the berm."[49] Mr. Lou Ullian (the Chief Range Safety Engineer) and his Range Safety people could take credit for encouraging the 45th Space Wing Commander to berm the blockhouse in November 1991. They also supported an initiative to build a soft, remote launch control center well outside the launch danger area. This led to the construction of the 1st Space Launch Squadron's Operations Building (a.k.a., the DELTA Operations Building) in 1995 and 1996, and the transfer of manned operations from the blockhouse to the new facility in 1997. Indeed, at a 45th Space Wing briefing in the Base Theater at Patrick AFB on 21 January 1997, Brigadier General Robert C. Hinson, the 45th Space Wing Commander, commended his people for their professionalism in handling the mishap and its aftermath. He complimented his Range Safety engineers in particular, since they had calculated blast damage and impact limit lines correctly. He also complimented the emergency response teams that responded to the fires at Complex 17.[50]

On the day of the accident, nine officials were chosen to serve on the AFSPC DELTA II Mishap Investigation Board in accordance with Air Force Instruction (AFI) 91-204. Colonel Ronald J. Haeckel was selected as Board President, and Colonel Craig S. Martin was chosen on 23 January 1997 to serve as Board Vice President.[51] The Board arrived at the Cape during the last week in January, and a team was established to assess damage and determine what repairs would be needed to return Complex 17 to launch-capable status. In the meantime, authorities banned all vehicles from parking inside the Flight Hazard Area (FHA) per General Hinson's letter of 23 January 1997. Alliant Techsystems, KSC Materials Division, and McDonnell Douglas assisted in the investigation to determine the cause of the accident. The analysis had to be extremely thorough – given the grave nature of the mishap – and the Board took its charter very seriously. Over the next two and one-half months, McDonnell Douglas evaluated 220 potential causes and completed 162 action items as part of the investigation. At the end of that time, McDonnell Douglas and the Board *were not able to determine conclusively what caused the SRM failure*, but a very short list

---

[48] Study, McDonnell Douglas, "Debris Study for DELTA II, GPS IIR-1, OPNR A4949 from Cape Canaveral AS, Pad 17A," ca 1 Mar 1997, p 21.

[49] The Debris Study also noted that "a direct hit on the blockhouse by one of the AI-GEMs (unpressurized solid rocket motors) could have resulted in serious injury or death." The blockhouse was operating under a safety waiver at the time of the launch. The blockhouse would not be manned for any subsequent launches because all blockhouse personnel were relocated to the new 1 SLS Operations Building shortly after the mishap

[50] Study, McDonnell Douglas, "Debris Study for DELTA II, GPS IIR-1, OPNR A4949 from Cape Canaveral AS, Pad 17A," ca 1 Mar 1997, p 13; Message, 45 SW/CP to HQ AFSPC/DOCP et al, "Incident Identification and Details/Subject: DELTA II Rocket/Operations Number A-4949," 302140Z Jan 1997; Memo, M. Cleary, 45 SW/HO, "Brigadier General Hinson briefed the Wing at the Base Theater...." 21 Jan 1997; 45 SW History, CY 1996, Vol I, p 19; 45 SW History, CY 1997, Vol I, pp 95, 97, 98.

[51] The Board's members included Lt. Colonel David J. Froiseth, Lt. Colonel Raymond E. Ebbs, Lt. Colonel James E. Gazur, Mr. John J. Erickson, Major Russell L. Porter, Major Ralph M. Strother, Captain Tarun K. Chattoraj, Captain Bradford E. Houser, and TSgt Charlie J. Powell.

of possible causes remained. Some of the failure mechanisms were less likely than others, so officials prepared the following short list of possible causes with that factor in mind:[52]

- Motor K-404 failed because its casing was weakened due to proof testing and associated water draining operations.

- Motor K-404 was damaged sometime during processing (but no evidence was uncovered to indicate this was likely).

- The motor casing was sabotaged (this was considered very unlikely).

Among its recommendations, McDonnell Douglas proposed the following actions:

- Inspect all SRMs (i.e., GEMs) ultrasonically to detect any critical delaminations or damage that might be introduced during manufacturing, proof testing or handling.

- Consider improving handling margins by increasing the hoop strength of motor cases used on future missions.

- Improve process control methods for resin, viscosity and Armalon application during motor case manufacturing operations.

- Consider reducing the upper hydro proof pressure and tolerance until motor case hoop strength is improved.

- Consider a program to periodically burst test a randomly selected motor case to verify the product's integrity.

- Modify SRM erection procedures at launch sites to minimize the possibility of overloading casings; improve SRM chock location, orientation and maintenance at all locations.

- Improve visual inspection procedures before cork and paint are applied to motor cases.

Later NAVSTAR GPS IIR Missions

The NAVSTAR IIR-1 flight turned out to be a very rare exception to the general rule for DELTA II operations. While a significant number of follow-on NAVSTAR IIR missions experienced setbacks over the next decade, none of them – as of this writing – failed. The next DELTA II mission, featuring the THOR II commercial communications spacecraft, was launched successfully from Pad 17A in late May 1997. The NAVSTAR IIR-2 mission followed, without incident, at 0343:00.579Z on 23 July 1997. The last Block IIA GPS spacecraft (NAVSTAR II-28)

---

[52] Special Order, HQ AFSPC, "SO M-1," 17 Jan 1997; Special Order , HQ AFSPC, "SO M-01," 23 Jan 1997; "Delta damage assessment still underway," *45th Space Wing Missileer*, 31 Jan 1997; Memo , Brigadier General Robert C. Hinson, 45 SW/CC, "Vehicle Parking within the Flight Hazard Area (FHA)," 23 Jan 1997; Letter , D.T. Sauer, Alliant Techsystems, to Ed Hikida, "Delta II Composite Analysis Test Summary," 19 Mar 1997; Excerpt , Director of Logistics Ops, Materials Science Division, Materials and Chemical Analysis Branch, LO-MSD-IM, "Failure Analysis of Graphite Epoxy Motor (GEM) Components From the U.S. Air Force Delta II Mission 241 (Global Positioning Satellite IIR-1) Rocket Launch Attempt From Space Launch Complex 17 (SLC-17) at Cape Canaveral Air Station (CCAS)," 17 Mar 1997; Report, McDonnell Douglas Aerospace, "DELTA 241 Failure Investigation Summary Report," 2 Apr 1997, pp 14, 22.

was launched from Pad 17A on 6 November 1997 at 0030:00.340Z.[53] That flight was successful, too.[54]

On the other hand, while most launch operations *might appear* smooth, controlled, predictable and unhurried on launch day, the lengthy process of getting DELTA IIs prepared to launch highly complex navigation satellites into space was a complicated and often frustrating undertaking. During the build-up for the GPS IIR-3 mission on Pad 17A in 1999, for example, preparations for the launch were delayed from 9 April through 3 May to accommodate the DELTA III ORION launch from Pad 17B on 5 May 1999. The GPS IIR-3 launch was rescheduled for 15 May 1999. As events unfolded, the GPS IIR-3 satellite remained in the Pad 17A White Room on 8 May 1999 when a severe thunderstorm passed over the launch site between 1400 and 1900 hours Eastern Daylight Time (EDT). Personnel evacuated Pad 17A around 1519 EDT, but the spacecraft had to be left behind, secured in the White Room.[55]

Unfortunately, the rain shielding in the White Room failed to protect the payload from water damage during the storm. According to the accident investigation report filed by Colonel Edwin E. Noble, 21st Logistics Group Commander, rain leaked through several openings in the White Room and pooled on top of the rain shield protecting the GPS IIR-3 spacecraft. The rain shield should have been taped on both sides, but it was only taped on one side. The added weight of the water finally collapsed the rain shield, and water leaked through hinge seams and quick release pinholes in the spacecraft's air-conditioning shroud hard cover. The water contaminated the spacecraft, and the original GPS IIR-3 satellite (SV-10) had to be removed and replaced with an entirely new satellite (SV-11).[56]

After technicians and engineers separated the original GPS IIR-3 spacecraft from the launch vehicle, the booster remained on Pad 17A until 27 May 1999. Then the contractor took it down to clear the launch pad for the DELTA II FUSE mission in late June 1999. The contractor re-erected the launch vehicle by the end of August, and officials completed the flight simulation on 8 September 1999. Hurricane FLOYD threatened the Florida coast about a week later, so protective measures were taken to minimize the effects of that 140-mile-per-hour storm. The contractor secured both of the Cape's DELTA II launch pads, locked the pads' Mobile Service Towers (MSTs) in place, and raised the towers off their wheels for stability. The GPS IIR-3 spacecraft was stored in its transportation canister inside the DSCS Processing Facility (DPF) until FLOYD passed 110 miles east of Cape Canaveral on 15 September 1999.[57]

Two days later, launch officials concluded there was no damage to the mission's flight hardware, and they approved a new launch date of 5 October 1999. Technicians and engineers mated the GPS IIR-3 spacecraft to the vehicle and installed the payload fairing on 30 September 1999. Lightning alerts stymied efforts to load propellant into the second stage on 2 October, but technicians fueled the second stage on 4 October. Officials readjusted the launch date to 6 October

---

[53] According to Captain Greg Wood and Lieutenant Eric Plott of the 1 SLS, the NAVSTAR II-28 satellite was placed in storage at Cape Canaveral in March 1993. The spacecraft weighed approximately 4,100 pounds, and it had an operational life expectancy of six and one-half years. The DELTA II selected for the II-2 mission was an „old timer.' It was delivered to the Cape in 1994 and placed in storage in 1995. The DELTA II was assembled on the launch pad as a pathfinder for the Avionics Upgrade program later on. It was used eventually as intended – to place a GPS spacecraft on-orbit.

[54] 45 SW History, CY 1997, Vol I, pp 99, 100, 101, 102.

[55] 45 SW History, CY 1999, Vol I, p 83.

[56] 45 SW History, CY 1999, Vol I, p 84.

[57] 45 SW History, CY 1999, Vol I, p 84.

1999 to compensate for the delay in fueling. Since the weather forecast for the 6th showed an 80 percent chance of weather violations during the launch window on that day, the launch was slipped to 7 October 1999. The countdown finally got underway on 7 October, and the DELTA II lifted off Pad 17A at 1251:00.686Z on 7 October 1999.[58]

Three more NAVSTAR IIR spacecraft were orbited successfully from Pad 17A in 2000. The first of them – GPS IIR-4 – was mated to its launch vehicle on 10 April 2000 as planned, but officials scrubbed the first launch attempt on 22 April 2000 at 2320Z due to problems with the spacecraft. They scrubbed the second attempt on 23 April at 2151Z. Once again, the scrub was called for spacecraft problems. Experts decided to take the next several days to analyze a technical issue involving electrical power supplied to the spacecraft. When the third and final countdown for the mission finally got underway on 10/11 May, a boat entered the launch area for a few minutes, rendering the range „Red' (i.e., not ready to support the launch). Despite that incident, there were no unplanned holds during that countdown, and the intruder cleared the danger area quickly. The DELTA II lifted off the launch pad at 0147:59.748Z on 11 May 2000.[59]

Launch vehicle problems delayed the next mission, but not seriously. Technicians and engineers erected the DELTA II booster for the GPS IIR-5 mission between 17 May and 10 June 2000. Second stage "troubleshooting" and an issue involving the third stage's rocket motor compelled officials to slip the launch from 21 June to 16 July 2000, but engineers mated the spacecraft to the vehicle without further delay on 5 July 2000. There were no weather hold-ups or other complications during the countdown on 16 July, and the range remained 'Green' (ready to support the launch) throughout the count. The DELTA II lifted off Pad 17A at 0917:00.450Z on 16 July 2000.[60]

Engineers and technicians erected IIR-6's launch vehicle on Pad 17A in early October 2000. High winds delayed GEM attachment operations on the 3rd, but technicians managed to complete the work three days later. Engineers mated the spacecraft to the vehicle as scheduled on 30 October 2000. A faulty Redundant Attitude Control System (RACS) module on the second stage failed during a retest on 30 October, but the module was replaced expeditiously. Technicians managed to complete the second stage propellant load as scheduled on 7 November. Officials scrubbed the first launch attempt on 9 November at 1525Z after an improperly wired B-Nut was discovered on a vernier engine's liquid oxygen line. The second countdown on 10 November was successful, and the DELTA II lifted off the launch pad as planned at 1714:02.219Z on 10 November 2000.[61]

Fourteen more NAVSTAR IIR replenishment satellites were launched from Cape Canaveral over the next eight years. (See Appendix C for launch dates, launch sites, and specific payloads associated with those missions, beginning with the GPS IIR-7 launch on 30 January 2001.) In many instances, technicians and engineers managed to erect a DELTA II on Pad 17A or Pad 17B and launch it well within 90 days, but there were some notable exceptions. The NAVSTAR IIR-8 mission,[62] for example, encountered very significant delays in 2002 after Range Safety analysts called the launch vehicle's Automatic Destruct System (ADS) into question.[63]

---

[58] 45 SW History, CY 1999, Vol I, pp 84, 85.

[59] 45 SW History, CY 2000, Vol I, pp vii, 109, 110.

[60] 45 SW History, CY 2000, Vol I, p 112.

[61] 45 SW History, CY 2000, Vol I, pp 113, 114.

[62] In addition to the NAVSTAR IIR-8 replenishment spacecraft, a three-foot-long, 65-pound USAF technology satellite – the XSS-10 Micro-satellite – was carried into orbit on the mission. Sponsored by the Air Force Research

29

Based on a 45th Space Wing safety recommendation in late January 2002, the 45th Space Wing Commander did not approve a safety waiver for the "non-ADS modified GPS IIR-8 mission" which was scheduled to lift off Pad 17B on 6 March 2002. A new ADS design had to be implemented before the mission was allowed to fly. As work on the ADS design continued, the launch slipped to 8 May 2002. Boeing's technicians delivered the DELTA II's first stage to the launch pad on 25 March 2002, but SMC's System Program Office (SPO) in Los Angeles directed Boeing to suspend all its GPS IIR-8, IIR-9, and IIR-10 launch processing activities on 27 March 2002.[64]

Colonel Michael T. Baker (the Detachment 8, SMC Commander) convened a DELTA II Technical Review Board (TRB) to investigate the matter and take corrective action. Boeing presented its TRB out-brief to Cape officials on 2 April 2002. Fifty-two of 59 action items identified by the Board during its investigation were closed out in early May 2002. The SMC Commander approved resumption of GPS IIR-8 processing a few days later. Technicians and engineers resumed processing on 15 May 2002, and officials approved a new launch date of 3 August 2002.[65]

Unfortunately, a DELTA IV Flight Termination System (FTS) wire harness failed during testing at Boeing's plant in Pueblo, Colorado, in early June 2002. Since the same type wire harness was installed on all DELTA II launch vehicles, erection of the DELTA II's second stage on Pad 17B was suspended until the harness failure could be resolved. The launch date was moved to 11 August 2002, but the lift-off soon slipped into indefinite status as the harness problem lingered. Officials eventually determined the machine used to process the failed wire harness had not damaged any DELTA II Flight Termination System (FTS) wire harnesses, so the matter was resolved on 20 September 2002. Technicians erected the DELTA II's second stage on 25 September 2002. Officials tentatively approved early November 2002 for the GPS IIR-8 launch, but other hardware problems soon arose to threaten further delays.[66]

Official conducted the simulation flight for GPS IIR-8 on 19 October 2002, but the third stage's spin table was damaged during a third stage/spacecraft lifting operation on 25 October. While officials debated how best to remove the payload safely, they noted the DELTA II third stage Star 48 motor's "one-year nozzle downtime limit" was due to expire in December 2002. Engineers and technicians removed the third stage/spacecraft from the DELTA II vehicle successfully on 8 November, and the payload canister was transported back to the DSCS Processing Facility and disconnected from the third stage on 15 November 2002. No damage to the spacecraft was detected, and the GPS satellite was placed in storage. The GPS IIR-8 Star 48 motor was shipped back to the Thiokol factory on 25 November 2002. In the meantime, officials decided to substitute the GPS IIR-21 spin table for the cracked spin table. The GPS IIR-21 spin

---

Laboratory (AFRL), the $100 million XSS-10 was designed to demonstrate line-of-sight guidance and inertial maneuvering capabilities needed for satellite maintenance, satellite inspection, and other duties. The AFRL's Space Vehicle Directorate at Kirkland AFB, New Mexico, monitored the 24-hour-long technology demonstration and commanded the micro-satellite's fiery reentry into the atmosphere at the end of the test.

[63] 45 SW History, CY 2003, Vol I, pp 93, 94.

[64] 45 SW History, CY 2003, Vol I, p 94.

[65] 45 SW History, CY 2003, Vol I, pp 94, 95.

[66] 45 SW History, CY 2003, Vol I, pp 95, 96.

table was dispatched from Pueblo, Colorado, and it arrived at the Cape on 22 November 2002. It was attached to the third stage shortly thereafter.[67]

By 3 December 2002, officials had approved a new launch date of 29 January 2003 for the GPS IIR-8 mission. The GPS IIR-8 Star 48 motor was recertified on 10 December, and it was returned to the Cape for reinstallation. Officials completed another simulation flight successfully on 17 December 2002. Due to high winds, the third stage/spacecraft mate was delayed until 18 January 2003, but it had no effect on the latest scheduled lift-off plan. Engineers mated the XSS-10 experimental payload to the DELTA II second stage on 20 January, and the payload fairing was installed on the vehicle on 24 January 2003. Following second stage propellant loading operations on 27 January, final preparations continued for the first launch attempt on 29 January 2003.[68]

The GPS IIR-8 mission was successful. Despite instrumentation problems on launch day, there were no unplanned holds during the countdown, and the DELTA II lifted off Pad 17B at 1806:00.431Z on 29 January 2003. Both payloads separated "as expected, without incident," thereby concluding a long and frequently delayed launch processing campaign.[69]

Spacecraft problems also delayed NAVSTAR IIR missions for several months in 2003 and 2005 (e.g., NAVSTAR IIR-10 and NAVSTAR IIR-14), but the long-term success of the NAVSTAR program was never in doubt. Beginning with the NAVSTAR IIR-14 (M) mission in late September 2005, the first of eight "modernized" IIR spacecraft were added to the constellation to enhance GPS capabilities.[70] The seventh in the series – GPS IIR-20 (M) – was launched successfully from Pad 17A at 0834:00.244Z on 24 March 2009. As of this writing, GPS IIR-21 (M) is scheduled to lift off Pad 17A in August 2009. Regarding future improvements in the NAVSTAR constellation, SMC awarded Lockheed Martin (and its associates ITT and General Dynamics) a $1,460,000,000 contract on 15 Mary 2008 to build the next-generation NAVSTAR III spacecraft. The DELTA II/NAVSTAR II launch effort remains one of the most remarkable accomplishments in the history of the Eastern Range.[71]

U.S. Military Technology Mission s

The first of three DELTA II missions dedicated solely to U.S. military technology experiments lifted off Complex 17 at 1615:00.626Z on 14 February 1990. It featured two Strategic Defense Initiative (SDI) payloads – LOSAT L and LOSAT R. The payloads were released in alphabetical order into two distinctly different circular orbits between 450 and 550 kilometers above Earth. The LOSAT L was controlled by the U.S. Naval Research Laboratory, headquartered

---

[67] 45 SW History, CY 2003, Vol I, pp 97, 98.

[68] 45 SW History, CY 2003, Vol I, p 98.

[69] 45 SW History, CY 2003, Vol I, p 99.

[70] After Lockheed Martin and ITT built 21 GPS IIR spacecraft, they subsequently modified eight of them (e.g., NAVSTAR IIR-14 through IIR-21) and added "(M)" to the designations of the ones that were modernized. Each of the upgraded satellites carried an improved antenna panel for greater signal power and two new military signals for better accuracy. The modernized spacecraft also had better encryption and anti-jamming capabilities than earlier IIR spacecraft.

[71] 45 SW History, CY 2003, Vol I, pp 114, 115, 116; 45 SW History, CY 2005, Vol I, pp 89, 90; 45 SW Command Post Checklist, "GPS IIR-20 Launch Report," 24 Mar 2009; E-Mail, 45 SW/CP to 30 SW/CP et al., "HOMELINE, 010, NOMINAL LAUNCH, 45 SW PATRICK AFB, FL," 24 Mar 2009; "Wing launches 20th GPS satellite," *45th Space Wing Missileer*, 27 Mar 2009; "Seventh Modernized GPS Satellite Built by Lockheed Martin Successfully Launched from Cape Canaveral," *Reuters*, 24 Mar 2009; Lockheed Martin, "Global Positioning System (GPS) – from Improved Accuracy to Knowing Exactly Where You Stand, There is One Important Word: How," undated.

in Bethesda, Maryland. The U.S. Air Force Weapons Laboratory in Albuquerque, New Mexico, was responsible for the LOSAT R system, but a control center in Hawaii controlled the LOSAT R spacecraft. The LOSAT L system was designed to measure the intensity of laser beams transmitted from the ground. The LOSAT R was designed to validate ground-based laser relay technology (e.g., beam stability, pointing, and beacon tracking).[72] The mission was successful.[73]

The second DELTA II military technology mission was launched in 2001. It featured the Geosynchronous Lightweight Technology Experiment (GeoLITE) sponsored by the National Reconnaissance Office (NRO). As the name suggests, the 4,000-pound GeoLITE spacecraft was injected into a geosynchronous transfer orbit during the flight. The spacecraft was built by TRW Defense Systems Division in partnership with the Hughes Space and Communications Company and MIT's Lincoln Laboratory. The experiment was designed to test two communications systems: 1) the Lightweight Technology Experiment - Lasercom Terminal, and 2) the Tactical Related Applications (TRAP) Data Dissemination System.[74]

The DELTA II's first stage was supposed to be erected Pad 17B on 9 March 2001, but the operation (and the launch date) slipped approximately two weeks when NRO authorities directed contractors to remove a repaired launch vehicle component and send it back to Rocketdyne for analysis. (The component passed with flying colors, and it was returned and reinstalled on the vehicle before the end of the month.) High winds delayed launch pad operations somewhat on 17 April and 8 May, but engineers and technicians managed to erect the DELTA II by 18 April and mate the payload to the launch vehicle on 9 May 2001. Technicians installed the payload fairing on 12 May. None of the delays in April or early May had any impact on the new 17 May 2001 launch date, but Rocketdyne uncovered leaks in two flex-hose lines on 16 May 2001. Technicians eventually replaced four flex-hose lines on the launch vehicle, but the countdown had to be slipped one day to allow workers time to complete the task. Officials rescheduled the launch for 18 May 2001.[75]

The DELTA II lifted off the launch pad at 1745:00.577Z on 18 May 2001. The initial boost phase was successful, and the second stage engine ceased firing approximately 11 minutes after lift-off. The vehicle and its payload coasted for awhile, and the subsequent firings of the second stage and third stage engines injected the payload into the proper geosynchronous transfer orbit. According to the 45th Range Squadron's post launch report, the GeoLITE launch was "the first time the USAF and the NRO bought a commercial launch service with the contractor responsible for all launch costs, [including] range and network [costs]." The flight also marked the first use of a DELTA II launch vehicle on an NRO mission.[76]

The third DELTA II military technology mission – the Micro-Satellite Technology Experiment (MiTEx) – was launched in late June 2006. The MiTEx was a joint technology concept demonstration program sponsored by the Defense Advanced Research Projects Agency (DARPA), the U.S. Air Force, and the U.S. Naval Research Laboratory to develop satellite and

---

[72] A third payload, LOSAT-X, was supposed to fly on this mission as well, but it was delayed due to technical difficulties. The LOSAT X payload was classified, so details concerning its features are not releasable. The payload eventually flew on the NAVSTAR II-11 mission on 4 July 1991.

[73] ESMC History, FY 1990, Vol I, pp 314, 315, 316.

[74] 45 SW History, CY 2001, Vol I, p 99.

[75] 45 SW History, CY 2001, Vol I, pp 99, 100.

[76] 45 SW History, CY 2001, Vol I, pp 100, 102.

launch technologies for rapid and responsive access to space. The MiTEx Space Vehicle (SV) consisted of an upper stage and two small satellites.[77]

In September 2005 officials scheduled the MiTEx launch for 28 February 2006, but hardware problems and a three-month-long machinists' union strike[78] delayed the mission. In addition, DELTA workers found some corrosion in one of their launch vehicles' second stage's oxidizer feed lines in late February, and the discovery prompted Aerojet to suspect similar contamination in all DELTA second stage oxidizer lines. The issue was resolved locally in late April 2006 by swapping the oxidizer flow meter on the MiTEx launch vehicle with the oxidizer flow meter on the GPS IIR-16 launch vehicle.[79]

Those issues aside, technicians and engineers erected the DELTA II on Pad 17A between 17 May and 24 May 2006, and range officials approved a new launch window, which eventually slipped one additional day to 21 June 2006. Hardware problems lingered into mid-June 2006, but engineers mated the payload to the DELTA II on 9 June, and technicians installed the payload fairing on 17 June 2006. The countdown went well, and the DELTA II lifted off Pad 17A at 2122:15.338Z on 21 June 2006. No safety actions were required, and the launch was successful.[80]

Foreign Military Missions

After Complex 17 was transferred back to the Air Force in 1988, DELTA IIs were used to launch foreign military spacecraft on three occasions. The DELTA IIs were clearly part of a brand new program, but the missions themselves continued a tradition of international cooperation dating back all the way to 1969, just a few years after NASA assumed responsibility for Complex 17.[81] Of the 10 foreign military missions launched from Complex 17 between November 1969 and December 1984, only one – SKYNET 2A – was unsuccessful. The NATO IV-A, NATO IV-B, and SKYNET 4D missions launched after 1988 continued the earlier tradition in the best sense of the word.[82]

The NATO III-D communications satellite was the last of the old Series III spacecraft. It was launched in November 1984 to serve as a „gap filler' satellite in the NATO constellation until enhanced NATO IV spacecraft could be introduced in later years. The NATO IV-A was the first of those enhanced satellites. It was launched from Pad 17B on a DELTA II Model 7925 vehicle in January 1991. The countdown on 7 January went smoothly for the most part, but a built-in hold had to be extended 68 minutes due to weather constraints. The vehicle lifted off the pad at

---

[77] 45 SW History, CY 2006, Vol I, p 74.

[78] The International Association of Machinists (IAM) Union went on strike on 2 November 2005, and DELTA operations at the Cape were refocused on essential caretaking to make sure environmental control systems on the various DELTA II and DELTA IV launch vehicles were maintained. The IAM strike ended in early February 2006, and IAM members returned to work around 10 February 2006.

[79] 45 SW History, CY 2006, Vol I, pp 74, 75.

[80] 45 SW History, CY 2006, Vol I, p 75.

[81] The NATO-A and NATO-B spacecraft were launched from Pad 17A in March 1970 and February 1971 respectively. The NATO III-A lifted off Pad 17B in April 1976, followed by the NATO III-C in November 1978. The NATO III-B and III-D missions were launched from Pad 17A in January 1977 and November 1984 respectively. The SKYNET A and SKYNET B missions were launched from Pad 17A in November 1969 and August 1970, and the SKYNET 2A and SKYNET 2B missions were launched from Pad 17B in January and November 1974 respectively.

[82] NASA Information Summary, "Major NASA Launches, Total Major ETR and WTR Launches," Dec 1989; 45 SW History Office, "Eastern Range Launch Database," updated 14 May 2009.

0053:01.060Z on 8 January 1991. The $110,000,000 NATO IV-A spacecraft entered its pre-planned 400 x 19,242-nautical-mile transfer orbit approximately half an hour later. As was true of earlier NATO communications satellites, the NATO IV-A was designed to provide communications between NATO member nations in Europe, the North Atlantic, and the eastern seaboard of the United States. The satellite had an on-orbit life expectancy of seven years.[83]

The NATO IV-B was launched successfully almost three years later. Matra-Marconi Space Systems built the spacecraft, which was assembled by British Aerospace Space Systems Ltd for the United Kingdom's Ministry of Defence. The satellite was designed to provide communications between NATO's command posts and military forces under their control.[84]

The second stage of the launch vehicle was off-loaded on 16 August 1993, and the first stage arrived at the Cape on 15 September 1993. Technicians and engineers erected the launch vehicle on Pad 17A between 3 November and 9 November 1993. The spacecraft was mated to the launch vehicle on 23 November, and – following beacon checks, engine preparations, and Class A ordnance work on 6 December 1993 – the DELTA II was readied for launch. Only one countdown was required, and the vehicle lifted off the launch pad at 0047:59.989Z on 8 December 1993.[85]

The third foreign military satellite to be launched aboard a DELTA II from Cape Canaveral was the 3,300-pound SKYNET 4D military communications spacecraft. Matra Marconi Space UK Limited produced it as the first of three new spacecraft to upgrade Britain's aging military communications satellite network. Technicians and engineers erected the launch vehicle on Pad 17B between 2 December and 16 December 1997, and they mated the spacecraft to the DELTA II on 2 January 1998. Payload fairing operations were completed on 6 January, and the second stage was fueled the following day.[86]

There were two unplanned holds during the countdown on 9 and 10 January 1998, but the first one cleared in less than one hour (for a vessel reported inside the "one ship contour") and the other one cleared within a few minutes. The DELTA II lifted off the launch pad at 0032:01.293Z on 10 January 1998. The launch was successful, and the spacecraft separated from the launch vehicle about an hour and a half later.[87]

---

[83] 45 SW History Office, "The Cape: Military Space Operations, 1971-1992," p 176; 45 SW History, 1 Oct 1990 – 31 Dec 1991, Vol I, pp 341, 342.

[84] 45 SW History, CY 1993, Vol I, p 219.

[85] 45 SW History, CY 1993, Vol I, pp 219, 220.

[86] 45 SW History, CY 1993, Vol I, pp 89, 90.

[87] 45 SW History, CY 1993, Vol I, p 90.

Clockwise from top left: 1) GPS IIR-19 spacecraft and its lift-off from Pad 17A on 15 Mar 2008; 2) 1 SLS Operations Building consoles and displays, 25 Mar 1997; 3) DELTA II GeoLITE launch, 18 May 2001; 4) a DELTA II first stage and interstage are erected on Complex 17, 26 Feb 1996; 5) Blockhouse 17 consoles and displays, 25 Oct 1996.

Clockwise from top left: 1) Complex 17 as it appeared on 28 Oct 1996; 2) the DELTA II carrying the GPS IIR-1 spacecraft explodes seconds after lift-off, 17 Jan 1997;
3) an aerial view of Complex 17; 4) Brig. Gen. Hinson (Center) and General Fogelman (2nd on Right) survey impact crater damage during the latter's visit, 14 Feb 1997;
5) 1 SLS Operations Building construction nears completion on 26 Feb 1996.

CHAPTER III

CIVIL MISSIONS

Beginning in June 1990, 30 DELTA IIs were launched from Complex 17 on a wide variety of purely scientific missions underwritten by NASA and other civilian space agencies. Though more modest in scope than the TITAN IV/Cassini mission and the Space Shuttle's Hubble Space Telescope, Compton Gamma Ray Observatory, and Chandra X-Ray Observatory missions launched in the 1990s, the DELTA II science flights of the past 20 years have followed in the footsteps of unmanned lunar, solar and interplanetary missions that began under NASA's leadership in the early 1960s. They also hearken back to the great maritime explorations of the 15th and 16th centuries when much of the world was unknown to everyone but the locals – yet ripe for discovery. As was true of those earlier adventures, the civil missions mentioned in this chapter would have been virtually impossible without the continued support of national institutions keenly interested in exploiting the capabilities of newly emerging technologies.

The DELTA II scientific missions can be divided into five categories: 1) astronomical surveys in various bandwidths beyond visible light, 2) solar wind observations and particle collections, 3) asteroid probes and comet encounters, 4) missions to Mars and Mercury, and 5) the Kepler Spacecraft „planetary search' mission. Most – but not all – of the missions were successful. Many took years to complete, so the summaries below explain how the missions progressed after they were launched. All of the missions were launched while the 1st Space Launch Squadron was responsible for Complex 17.

X-Ray, Gamma Ray, Infrared, Microwave Anisotropy, and Extreme Ultraviolet Astronomy Missions

A DELTA II carrying the Roentgen Satellite (ROSAT) was launched successfully from Pad 17A at 2147:59.850Z on 1 June 1990. Sponsored as a cooperative mission between West Germany's Ministry for Research and Technology, NASA and the United Kingdom, the ROSAT was designed to investigate black holes (i.e., gravity wells) and study x-rays emitted from stars and remnants of supernovas. West Germany had exclusive use of the spacecraft during the first six months of the ROSAT's operation, but NASA and the United Kingdom were allowed shared access subsequently. NASA reimbursed the Air Force for procuring the commercial DELTA II that was used to carry out the ROSAT into space.[1]

The DELTA II was launched on a flight azimuth of 65 degrees. The flight plan required two "dogleg" maneuvers during the first two minutes of the mission to minimize the launch vehicle's angle of attack. Another dogleg maneuver was initiated approximately four and one-half minutes after lift-off to give the payload an orbital inclination of 53 degrees. The spacecraft was released approximately 44 minutes after the launch, and it entered its pre-planned 580-kilometer circular orbit successfully. After more than four years of operation, one of the ROSAT's major instruments – the Position Sensitive Proportional Counter – was turned off to save its remaining supply of detector gas for later observations. The counter was used again in

---

[1] ESMC History, FY 1990, Vol I, p 317; News Release, McDonnell Douglas, "ROSAT DELTA Launch 195," 11 May 1990.

conjunction with the ROSAT's x-ray telescope to complete an all-sky survey in 1997 and for several other observations in 1998 and early 1999. The ROSAT was turned off on 12 February 1999.[2]

Two years after the ROSAT was launched, another DELTA II lifted off Pad 17A to carry NASA's Extreme Ultraviolet Explorer (EUVE) into low-Earth orbit. The EUVE mission was a major milestone in the history of high-tech astronomy. It was designed to open what scientists termed "the last unexplored spectral window in astrophysics." It was NASA's first dedicated extreme ultraviolet mission.[3] Under NASA's sponsorship, the Space Sciences Laboratory in Berkeley, California, was able to design and build the EUVE to observe and collect EUV data in the 70–to-760 Angstrom energy range. The result was a 7,000-pound spacecraft equipped with three scanning telescopes to conduct an "all-sky, all-bands" survey and an EUV spectrometer to detect objects outside our galaxy.[4]

Officials scrubbed the first EUVE countdown for bad weather on 6 June 1992, but the second countdown led to a successful lift-off at 1639:59.786Z on the following day. The spacecraft separated from the launch vehicle approximately one hour after lift-off, and the EUVE entered in a 292-nautical-mile-high circular orbit. The spacecraft and its instruments were checked out thoroughly before they went into operation about a month later. The spacecraft's scanning telescopes mapped the entire sky in EUV mode over the next six months, and a deep survey was performed. Images of more 800 objects were captured and cataloged during the all-sky survey. The mission also featured the first EUV detection of objects outside our galaxy.[5]

The EUVE remained operational for eight years, but its orbit decayed to the point of reentry in early 2002. It finally reentered Earth's atmosphere on January 30, 2002. U.S. Space Command radar covered the event. A few small pieces of the spacecraft may have fallen on Egypt, but Egyptian authorities denied they detected any debris from the spacecraft's uncontrolled reentry. The mission was highly successful.[6]

NASA's next DELTA II-boosted x-ray astronomy mission – the Rossi X-Ray Timing Explorer (RXTE) – was launched from Pad 17A at 1348:00.125Z on 30 December 1995.[7] The

---

[2] ESMC History, FY 1990, Vol I, p 318; Fact Sheet, Goddard Space Flight Center, "The ROSAT Mission," 11 Jan 1995; Goddard Space Flight Center, "The ROSAT Mission (1990 – 1999)," 7 Sep 2008.

[3] Scientists made their first EUV observations of the Sun in 1959, but they had to wait more than 16 years for instruments capable of detecting and recording EUV sources outside our solar system. The delay was due partially to the limited sensitivity of standard telescopic mirrors, but inventors solved that handicap by developing precision-made 'grazing incidence' mirrors that collected EUV light when the mirrors were almost parallel to the incoming light source. Scientists at the University of California, Berkeley, developed detectors without protective covers. Lacking covers, the instruments were so sensitive they were able to pick up incoming photons and record them the instant they were received. Diffraction gratings were used to separate the radiation into individual wavelengths.

[4] Space Sciences Laboratory, University of California, Berkeley, "All About the Extreme Ultraviolet Explorer," 21 Mar 2001.

[5] 45 SW History, CY 1992, Vol I, p 234; Space Sciences Laboratory, University of California, Berkeley, "All About the Extreme Ultraviolet Explorer," 21 Mar 2001.

[6] Goddard Space Flight Center, "EUVE: New Postings – EUVE spacecraft re-enters Earth's atmosphere," posted 31 Jan 2002; "Satellite's debris, if any, probably fell on Egypt," *USA Today,* 31 Jan 2002.

[7] Unfortunately, the mission was plagued with launch scrubs during the three weeks leading up to final lift-off. Officials scrubbed the first launch attempt at 1555Z on 10 December due to upper and lower wind conditions. Wind

Goddard Space Flight Center built, managed and controlled the RXTE, and the spacecraft carried three instruments[8] to gather data on a variety of x-ray sources (e.g., "white dwarf" stars, accreting neutron stars, black holes and active galactic nuclei). The DELTA II Model 7925 launch vehicle chosen for the mission was the first operational launch vehicle equipped with McDonnell Douglas' new Redundant Inertial Flight Control Assembly (RIFCA) guidance system. The RXTE was launched into a 362-mile-high orbit inclined 23 degrees to the equator.[9]

NASA officials expected the RXTE to operate for as long as five years on-orbit, but the spacecraft surpassed all expectations. With the RXTE's latest operational cycle due to end in December 2008, NASA decided to extend RXTE operations through September 2009. Despite the difficulties encountered in getting the launch vehicle assembled and launched in 1995, the mission turned out to be a great success.[10]

The second DELTA II mission to feature an ultraviolet astronomy payload lifted off Pad 17A on 24 June 1999.[11] The spacecraft was the Far Ultraviolet Spectroscopic Explorer (FUSE), and it represented the culmination of an international effort involving NASA, the Canadian Space Agency, and the France's Centre National d'Etudes Spatiales. The spacecraft was designed to gather high-resolution data on faint objects throughout the Milky Way and beyond the galaxy in the 910 to 1180 Angstrom range. The University of Colorado built the FUSE's spectrograph. Johns Hopkins University built the FUSE's telescope and managed the mission. NASA expected the satellite to provide a wealth of information on the density, pressure, temperature and composition of plasmas and gases. The object of the flight was to place the FUSE spacecraft into a circular 418.5 nautical-mile-high Earth orbit inclined 25 degrees.[12]

Boeing selected a DELTA II Model 7320-10 vehicle equipped with three GEMs for the mission. Engineers and technicians erected the vehicle between 3 and 8 June 1999. The spacecraft was mated to the DELTA II on 16 June, and second stage fueling operations were completed on 22 June 1999. Shortly after lift-off on the 24th, the vehicle rolled into a flight azimuth of 97.5 degrees. It achieved initial low-Earth orbit approximately 10 minutes later. A "slow-roll" maneuver provided thermal conditioning for the vehicle and payload before the DELTA II's second stage restarted around one hour and seven minutes into the flight. A brief

---

conditions also forced similar scrubs on 11, 12 and 17 December 1995. The wind cooperated during the countdown on 18 December, but the DELTA II shut itself down prior to engine start on the 18th because its main oxidizer valve had frozen shut due to moisture accumulation from the previous week. Technicians purged, cleaned, reassembled and tested the system, and the vehicle was prepared (once again) for a launch attempt on 29 December 1995. Upper level winds forced officials to scrub the launch at 1401Z on 29 December, but the countdown on 30 December final led to a successful launch.

[8] The instruments were: 1) the High Energy X-Ray Timing Equipment (HEXTE), 2) the All-Sky Monitor (ASM), and 3) the Proportional Counter Array (PCA).

[9] 45 SW History, CY 1995, Vol I, pp 110, 111.

[10] Fact Sheet, Goddard Space Flight Center, "Rossi X-Ray Timing Explorer Mission (1995-present)," undated.

[11] Only one countdown was required to launch the FUSE mission. The range remained 'Green' until 1529Z when a small boat was detected approaching the one-boat contour box. Officials extended the final built-in hold for about five minutes so the area could be cleared and the boat could be turned away from the danger area. The countdown resumed at 1540Z, and the DELTA II lifted off the launch pad at 1543:59.879Z on 24 June 1999.

[12] 45 SW History, CY 1999, Vol I, pp 79, 80; News Archive, NASA, "NASA Concludes Successful Fuse Mission," 17 Oct 07.

restart of the second stage engine ended the DELTA II's powered flight. The spacecraft separated from the upper stage about 10 minutes later, and the FUSE entered its 418.5 nautical-mile-high circular orbit.[13]

The FUSE's three-year-long orbital science mission got underway on 1 December 1999 as planned. Operations continued until 10 December 2001 when the satellite went into "safe mode" and shut down after the second of its four reaction wheels (used to steady and point the spacecraft) quit spinning. Engineers and scientists at Johns Hopkins and the Goddard Space Flight Center joined forces with contractors working for Orbital Sciences Corporation and Honeywell Technology Solutions, Inc., and together they managed to overcome the problem. The FUSE observatory was back in operation by early March 2002. In the meantime, NASA recommended extending the mission to five full years. FUSE enjoyed an eight-year-long run, ending its operations in July 2007. NASA was very pleased with the highly successful mission, and the agency touted it for giving "astronomers a completely new perspective on the Universe."[14]

An ambitious effort to map the cosmos' microwave background radiation led to NASA's next DELTA II astronomy mission – the Microwave Anisotropy Probe (MAP).[15] Boeing selected a DELTA II Model 7425 launch vehicle equipped with four GEMs for the mission. Engineers and technicians erected the vehicle on the launch pad between 24 May and 30 May 2001, and the spacecraft was mated to the launch vehicle on 19 June 2001. Following lift-off from Pad 17B at 1946:46.183Z on 30 June 2001, the vehicle pitched into a flight azimuth of 95 degrees. The second stage engine ceased firing approximately 12 minutes later, and a pitch maneuver reoriented the vehicle and spacecraft for second stage restart and third stage burn. The vehicle and its payload coasted for approximately one hour. Subsequent firings of the second stage and third stage engines injected the payload into the proper orbit approximately one hour and 25 minutes after lift-off.[16]

The MAP spacecraft took three months to reach the second Lagrange point – L2 (e.g., about one million miles farther out from the Sun than Earth). Once there, the spacecraft spent six months surveying the sky. Researchers needed an additional nine months to check the MAP's data, so the first useful data from the mission was not available until 18 months after the launch. Nevertheless, the MAP completed its second full year of operations in September 2003. The MAP team released a set of papers and flight data covering the first year of observations in February 2003. The team released four more years' worth of information by March 2008, and

---

[13] 45 SW History, CY 1999, Vol I, pp 79, 80.

[14] "Far-Ultraviolet Spectroscopic Explorer Orbiting Observatory Returns to Full-Time Science Operations," *ScienceDaily*, 7 Mar 02; News Archive, NASA, "NASA Concludes Successful Fuse Mission," 17 Oct 07.

[15] The Goddard Spaceflight Center and Princeton University sponsored the MAP observatory. The spacecraft measured 3.8 x 5.0 meters, and it weighed approximately 1,850 pounds. It was designed to measure temperature fluctuations (a.k.a. anisotropy) and to produce a highly detailed map of cosmic microwave background radiation from 22 to 90 GHz "over the entire sky." The MAP's sponsors hoped the results would give scientists new insights into the formation of stars and galaxies. The observatory's solar arrays produced 419 watts of power and shielded the spacecraft from thermal damage. The spacecraft cost $95 million, and the launch service cost $50 million. In addition to those costs, $14 million was budgeted to cover two years' worth of MAP data gathering and mapping operations.

[16] 45 SW History, CY 2001, Vol I, pp 102, 103.

that material provided a detailed look to the early years of the Universe. As of this writing, the mission has been extended twice, and it is not scheduled to end until at least September 2009.[17]

The object of NASA's next DELTA II astronomy mission was to inject the agency's Space Infrared Telescope Facility (SIRTF) Observatory into a rather unusual orbit which would allow the spacecraft to trail behind Earth as it orbited the Sun.[18] The SIRTF Observatory was touted as "the last of NASA's orbiting Great Observatories" — a family of spacecraft that included the Hubble Space Telescope, the Chandra X-Ray Observatory, and the Compton Gamma Ray Observatory – all launched under NASA's sponsorship in the 1990s.[19] The SIRTF Observatory was designed to study a wide range of objects stretching from our solar system to the "distant reaches" of the Universe. According to the final revised flight scenario, the SIRTF launch could occur during one "instantaneous" launch window available daily from late August through the end of September 2003.[20] This special mission required a special vehicle – the DELTA II 7920 Heavy (7920H) – equipped with significantly larger Alliant GEMs[21] than the GEMs flown on standard DELTA II Model 7925 launch vehicles.[22]

NASA planned to launch the SIRTF mission on 29 March 2003 initially, but the path to Pad 17B was eventful. Changes in the schedule soon forced officials to slip the launch to mid-April 2003. Technicians finished mating the GEMs to the launch vehicle on 10 March, and the DELTA II's second stage was erected on 14 March 2003. Unfortunately, two GEMs displayed nozzle delaminations that NASA accepted initially but later rejected. The launch fell into indefinite status on 14 April, and Boeing's problems were compounded when technicians discovered the aft fairings on the GEMs did not fit properly. In view of those problems, NASA decided to delay the mission. The SIRTF payload was removed from the booster on Pad 17B, and the DELTA II core vehicle was reconfigured to support the MARS EXPLORATION ROVER-B (MER-B) mission in July 2003. NASA requested a new launch date of 23 August 2003 for the SIRTF mission, and it was approved on 2 June 2003.[23]

Following the MER-B launch on 8 July, NASA and Boeing refocused their attention on the SIRTF mission. Engineers erected a new DELTA II Heavy booster on Pad 17B between 18

---

[17] Fact Sheet, Goddard Space Flight Center, "WMAP Facts," updated 14 October 2008; "Mapping Solutions To Cosmic Riddles," *Aviation Week & Space Technology*, 25 Jun 2001.

[18] Once the spacecraft escaped Earth's gravitation, it was injected into a heliocentric orbit to drift slowly away for Earth while maintaining approximately the same distance from the Sun (e.g., 93 million miles). By early May 2009, the SIRTF was trailing our planet by approximately 62 million miles.

[19] The Space Shuttle *Discovery* placed the Hubble Space Telescope on-orbit in April 1990. The Compton Gamma Ray Observatory was released on-orbit by the Space Shuttle *Atlantis* in April 1991. The Space Shuttle *Columbia* orbited the Chandra X-Ray Observatory in July 1999.

[20] If the DELTA II was launched before 27 August, the flight azimuth had to be 105 degrees. If it was launched on or after 27 August, the flight azimuth became 107 degrees.

[21] The DELTA II Heavy's GEMs were four feet longer and six inches wider than standard GEMs. They provided approximately 140,000 pounds of thrust each. Owing to their greater size, the Heavy's GEMs carried more fuel and burned 13 seconds longer than standard GEMs.

[22] 45 SW History, CY 2003, Vol I, pp 109, 110; NASA Facts, JPL/NASA, "Space Infrared Telescope Facility," 20 Dec 2002.

[23] 45 SW History, CY 2003, Vol I, pp 110, 111.

July and 29 July 2003. Engineers mated the SIRTF payload to the launch vehicle on 10 August, and technicians installed the payload fairing on 14 August 2003. The tardy arrival of the support ship OTTR prompted NASA to slip the SIRTF launch from 23 to 25 August 2003. There were no unplanned holds during the countdown, and the DELTA II lifted off Pad 17B without incident at 0535:39.231Z on 25 August 2003. The flight marked the 300th launch in the history of the DELTA program.[24]

The SIRTF was renamed the Spitzer Space Telescope in December 2003 in honor of Lyman Spitzer, a prominent 20th Century astronomer. Under either name, the spacecraft provided almost six years of service before the liquid helium that kept its infrared sensors chilled to minus 456 degrees Fahrenheit ran out in mid-May 2009. While that event closed out one phase of Spitzer's mission, it did not mark the end of the telescope's usefulness. Adapting to a slightly warmer environment of minus 404 degrees Fahrenheit, two of the telescope's shortest-wavelength detectors were recalibrated by scientists to continue the mission thereafter.[25]

Spitzer has rewarded its makers generously over the years, providing enormous amounts of data on comets, stellar "nests," and how stars are born. In 2005, Spitzer detected the first photons from a planet orbiting around another star. Spitzer also detected hundreds of black holes while observing galaxies billions of light years away. Astronomical data gathered from the project has been cited in more than 1,500 scientific papers, and more treatises will be written in the near future.[26]

Gamma rays were the subject of NASA's next DELTA II astronomical mission, which involved injecting a 3,234-pound astronomical observatory named SWIFT into a 324-nautical-mile-high circular orbit in late November 2004. SWIFT was designed to detect, observe and characterize approximately 200 gamma ray bursts based on x-ray afterglow from the bursts. The spacecraft was equipped with a Burst Alert Telescope (BAT), an X-Ray Telescope (XRT), and an Ultraviolet/Optical Telescope (UVOT) to gather data. One of the major objectives of the mission was to detect gamma ray bursts from extremely distant, first-generation stars. NASA hoped SWIFT would give astronomers their most complete set of data on the Universe's "mysterious" gamma ray explosions. SWIFT would also provide an "all-sky" survey in high-energy x-ray wavelengths.[27]

General Dynamics Spectrum Astro built the SWIFT spacecraft, and the DELTA II chosen for the flight was equipped with three GEMs. NASA contributed $232 million for the $250 million mission. The Italian Space Agency and the United Kingdom Particle Physics and Astronomy Research Council provided additional funding.[28]

Engineers and technicians were supposed to erect the DELTA II's first stage on Pad 17A on 1 September 2004, but the approach of Hurricane FRANCES delayed the operation. A second attempt to raise the booster was foiled by the approach of Hurricane JEANNE in late September.

---

[24] 45 SW History, CY 2003, Vol I, p 111, 112.

[25] Whitney Clavin, "If Spitzer Could Talk: An Interview with NASA's Coolest Space Telescope," 5 May 2009; NASA, "NASA's Spitzer Begins Warm Mission – Spitzer Space Telescope Mission Status," 15 May 2009.

[26] JPL/NASA, "Spitzer Telescope Warms Up to New Career," 6 May 2009.

[27] 45 SW History, CY 2004, Vol I, pp 104, 105.

[28] 45 SW History, CY 2004, Vol I, p 105.

Officials rescheduled the launch from 26 October to 8 November 2004 to accommodate those delays. Engineers erected the second stage on 8 October 2004, but a recent "stand alone" failure involving a Redundant Inertial Flight Control Assembly (RIFCA) on a weather satellite mission had implications for any DELTA II featuring a RIFCA.[29] Officials placed the SWIFT launch in indefinite status on 25 October, though in that instance the action was taken to make way for the DELTA II GPS IIR-13 mission slated to launch in early November 2004. Boeing requested a new launch date for SWIFT, and the 45th Space Wing ultimately chose 20 November 2004 for the lift-off. The payload was mated to the launch vehicle on 8 November 2004.[30]

There was one extended hold and one unplanned hold during the countdown on 20 November, but the launch went well. The DELTA II lifted off Pad 17A at 1716:00.661Z on 20 November 2004. Separation occurred as planned, and a control team at Pennsylvania State University began checking out the spacecraft once it was on-orbit.[31]

The Goddard Space Flight Center continues to manage the SWIFT mission – which is ongoing – and the spacecraft is operated in concert with Pennsylvania State University, Los Alamos National Laboratory, and General Dynamics. The project's international partners include the University of Leicester and Mullard Space Sciences Laboratory in Great Britain, the Italian Space Agency, and Italy's Brera Observatory. In late April 2009, SWIFT detected the gamma ray burst of a star that died when the Universe was only 630 million years old. As of this writing, it remains the most distant (and earliest) cosmic explosion on record.[32]

The Cape's next – and most recent – DELTA II gamma ray mission featured the Gamma Ray Large Area Space Telescope (GLAST). Sponsored by NASA, the U.S. Department of Energy, and institutions in Germany, France, Japan, Italy and Sweden, the GLAST was designed to collect data on the widest range of gamma ray sources (e.g. neutron stars, black holes, supernovas and pulsars) with state-of-the-art technology.[33] Touted as a "next generation" high energy gamma ray observatory, GLAST was supposed to be launched on 7 October 2007. Unfortunately, the mission slipped into indefinite status in late April 2007, and hardware problems and technical constraints delayed the launch until 11 June 2008.[34]

---

[29] NASA decided in late October 2004 to have the contractor remove the SWIFT RIFCA and replace it with another RIFCA that had been x-rayed onsite. The new RIFCA was tested with good results on 10 November 2004.

[30] 45 SW History, CY 2004, Vol I, pp 105, 106.

[31] 45 SW History, CY 2004, Vol I, p 106.

[32] NASA, "New Gamma-Ray Burst Smashes Cosmic Distance Record," 28 Apr 2009.

[33] The GLAST spacecraft was built by General Dynamics in California, and it was transported to the U.S. Naval Research Laboratory in Washington D.C. to ensure it could withstand the rigors of spaceflight and operations on-orbit. It featured a Large Area Telescope (LAT) equipped with detectors to convert incoming gamma rays into electrons ranging from 20 MeV to 300 GeV. It also had a GLAST Burst Module to augment detection between 10 KeV and 20 MeV in strength.

[34] 45 SW History, CY 2008, Vol I, pp 117, 118; Summary, Goddard Space Flight Center, "GLAST, The Gamma-ray, Large Area Space Telescope," *gsfc.nasa.gov.,* 20 Jun 2008; Summary, "Gamma-ray Large Space Telescope," *aerospaceguide.net.,* undated; Summary, Goddard Space Flight Center, "GLAST Overview," *gsfc.nasa.gov.,* 12 Aug 2008; News Release, Marshall Space Flight Center, "New gamma ray burst instrument powers up," 30 Jun 2008.

The GLAST spacecraft weighed 9,429 pounds at lift-off, so it required a DELTA II 7920H (Heavy) launch vehicle to get into low-Earth orbit.[35] The DELTA II lifted off Pad 17B at 1605:00.512Z on 11 June as scheduled, and it rolled into a flight azimuth of 94 degrees. Following GEM separation, the DELTA II's main engine quit thrusting about four and one-half minutes into the flight. The second stage burn followed a few seconds after first stage/second stage separation, and the 10-foot diameter payload fairing covering the second stage and spacecraft was jettisoned early in the second stage's first burn. Following a lengthy coasting period and several second stage restarts, the spacecraft was placed in a 350-mile-high circular orbit as planned.[36]

Two weeks later, scientists gathered at NASA's GLAST Burst Monitor Instrument Operations Center in Huntsville, Alabama, to observe the activation of the GLAST Burst Module (GBM). All 14 GBM detectors were activated successfully at 7:45 p.m. Central Daylight Time on 25 June 2008. The event confirmed the launch had been successful, and it became an important milestone in the GLAST's on-orbit preparation for full operations.[37]

Solar Wind Missions

Six DELTA II solar wind missions were launched from Complex 17 between 24 July 1992 and 18 February 2007. The first – GEOTAIL/DUVE – was a collaborative effort by the Institute of Space and Astronautical Science (ISAS) and NASA. The cylindrical-bodied GEOTAIL was approximately 1.6 meters long and 2.2 meters in diameter, and it weighed approximately 2,200 pounds at lift-off. The spacecraft was designed by the Japanese Institute of Space and Astronautical Science to explore solar wind plasma within the geomagnetic tail (hence GEOTAIL) of Earth. The GEOTAIL spacecraft was the principal payload, but the Diffuse Ultraviolet Explorer (DUVE) was included as a small payload on the DELTA II's second stage.[38] The GEOTAIL mission involved orbiting the spacecraft for two years in a double-lunar swing-by orbit, followed by one year of observations at closer range to study magnetotail substorms near Earth.[39]

The countdown on 24 July 1992 went smoothly, and the DELTA II lifted off Pad 17A at 1426:00.015Z. The mission was successful, and officials continued to vary the GEOTAIL's orbit well into 1997 to collect additional data. Researchers detected important clues in the "plasma sheet," which was the source of aurora electrons. When amplified with data gathered by other

---

[35] The DELTA II Heavy was equipped with nine longer-than-standard Alliant GEMs and a second stage that featured an Aerojet AJ10-118K Improved Transtage Injector Program (ITIP) engine.

[36] 45 SW History, CY 2008, Vol I, pp 118, 119.

[37] News Release, Marshall Space Flight Center, "New gamma ray burst instrument powers up," 30 Jun 2008; "GLAST Launch," *45th Space Wing Missileer,* p 1, 13 Jun 2008; 45 SW History, CY 2008, Vol I , p 119.

[38] The DUVE featured a far ultraviolet imaging spectrometer with an electronic storage module linked to the launch vehicle's second stage telemetry system. The DUVE mission was designed to detect hot interstellar gases and relay that data to ground operators. Data collection began after the second stage and DUVE separated from the upper stage and GEOTAIL in flight, and it continued for approximately eight hours while the DUVE and second stage remained in Earth orbit.

[39] 45 SW History, CY 1991, Vol I, p 235; Institute of Space and Astronautical Science, "GEOTAIL Mission Profile," undated; Goddard Space Flight Center, "GEOTAIL Project Overview," undated.

nations' satellites, NASA uncovered new facts related to the timing of magnetic reconnection (a magnetic energy release effect).[40]

The next three solar missions – WIND, ACE, and GENESIS – were equally encouraging and mostly successful.

The WIND mission was launched from Pad 17B at 0931:00.061Z on 1 November 1994. Martin Marietta Astro Space built the WIND spacecraft, and NASA managed the project for the Global Geospace Science (GGS) program. McDonnell Douglas selected a DELTA II Model 7925-10 launch vehicle equipped with nine GEMs to fly the mission, which was part of the International Solar Terrestrial Physics (ISTP) initiative. WIND was designed to measure the solar wind phenomena believed responsible for magnetospheric dynamics. The WIND spacecraft was 2.4 meters in diameter and 1.8 meters high. It carried 300 kilograms of propellant, which raised its total weight to approximately 1,250 kilograms at lift-off. WIND made its observations from a position near the Lagrange point L1 initially, but its orbit was changed to sample other areas of space surrounding Earth after the Solar and Heliospheric Observatory (SOHO) was placed in a similar (L1) orbit in December 1995. With an operational lifespan of at least three years, WIND gathered data that complemented and reinforced data collected by other „solar wind' spacecraft.[41]

By the end of 1997, WIND had rounded the L1 Lagrange point on a heading back toward Earth, but the spacecraft's useful life was just entering a new chapter. The Moon's gravity was used to alter WIND's trajectory into a set of "petal" orbits that would allow the spacecraft to explore the magneto sheath for several years. The ISTP project ended in October 2001 with the WIND spacecraft in a supporting role for other solar wind research projects around the world.[42]

A DELTA II carrying the Advanced Composition Explorer (ACE) lifted off Pad 17A at 1439:00.459Z on 25 August 1997. The launch was just the first leg in the spacecraft's two-to-five-year mission between Earth and the Sun. The primary objective of the ACE mission was to determine and compare the composition of several samples of matter (e.g., solar corona, interplanetary medium, local interstellar medium, and galactic matter) gathered along a modified halo orbit around the Earth-Sun libration point. Nine science instruments were used to capture the required data. The Goddard Space Flight Center managed the ACE spacecraft, and the Johns Hopkins University Applied Physics Laboratory and its support contractors designed and integrated the payload.[43]

The ACE spacecraft achieved orbit at the L1 libration point (i.e., where the gravitational pull from Earth just matches that of the Sun) as planned. The mission was highly successful. As of this writing, more than 400 peer-reviewed scientific papers have been published by members of the ACE science team based on ACE data. With the exception of the Solar Energetic Particle Ionic Charge Analyzer (SEPICA), the spacecraft's instruments were still in good working order

---

[40] 45 SW History, CY 1991, Vol I, pp 235, 236; Institute of Space and Astronautical Science, "GEOTAIL Mission Profile," undated.

[41] 45 SW History, CY 1994, Vol I, pp 109, 110; Fact Sheet, NASA, "WIND Mission," Jan 1994; Fact Sheet, Martin Marietta Astro Space, "GGS WIND," 1994; Summary, GSFC, "WIND," updated 25 Nov 2001, p 1.

[42] Fact Sheet, NASA Goddard Space Spaceflight Center, "WIND," updated 25 Nov 2001.

[43] 45 SW History, CY 1997, Vol I, pp 100, 101.

in the spring of 2008. The spacecraft contained sufficient propellant to maintain its prescribed orbit until approximately 2024.[44]

On 8 August 2001, a DELTA II lifted off Pad 17A at 1613:40.324Z with NASA's GENESIS spacecraft onboard. The object of that flight was to inject the payload into a halo orbit around the Sun.[45] Lockheed Martin built the spacecraft, and NASA's Jet Propulsion Laboratory (JPL) managed the program. Boeing selected a DELTA II 7326 vehicle equipped with three GEMs for the mission. According to the flight scenario, the DELTA II lifted off the pad and pitched into a flight azimuth of 95 degrees. The second stage engine ceased firing approximately 10 minutes and 30 seconds after lift-off. A pitch maneuver reoriented the vehicle and spacecraft for second stage restart and third stage burn. The vehicle and its payload coasted for approximately 45 minutes. Subsequent firings of the second stage and third stage engines injected the payload into the proper orbit. The spacecraft separated from the vehicle approximately one hour after lift-off. The GENESIS spacecraft arrived at the first Lagrange point (e.g., about one million miles from Earth and 92 million miles from the Sun) three months later.[46]

The mission went extremely well until the sample return canister came back to Earth. On 8 September 2004, the canister's drogue chute and parafoil both failed to deploy, and the GENESIS capsule hit the Utah Test and Training Range at 193 miles per hour. The sample container was breached on impact, but three out of four segments containing oxygen and nitrogen isotopes were intact. In March 2005, NASA announced that mission researchers had identified ions from the solar wind in one of the wafer fragments contained in the canister. The $264 million GENESIS mission was, at the very least, partially successful.[47]

The next DELTA II solar mission was launched from Pad 17B at 0052:00.339Z on 26 October 2006. The object of that flight was to place the Solar TErrestrial RElations Observatory (STEREO) payload into a 90-nautical-mile circular parking orbit, followed by a coasting period and another boost (via second and third stage burns) to the orbit required for payload separation. Once STEREO was deployed in heliocentric orbit, it operated as *two* separate, golf cart-sized

---

[44] Fact Sheet, Eric R. Christian and Andrew J. Davis, University of New Hampshire Space Science Center, "Advanced Composition Explorer (ACE) Mission Overview," 15 Apr 2008; Summary, UNH Space Science Center, "SEPICA in Brief," undated.

[45] The GENESIS spacecraft was a relatively low cost, multiple sample return collector of solar matter. It was designed to collect particles from the solar wind (e.g., isotopes of oxygen, nitrogen, the noble gases and other elements) and return them to Earth. GENESIS was equipped with a contamination-proof canister for that purpose, and the spacecraft's collector arrays would be exposed to the solar wind for two years to collect the samples. The spacecraft needed three months to reach the first Lagrange point (L1), two years on-station to collect the samples, and five months to get the samples back to Earth. If the mission ended according to plan, the samples would be recovered over the Utah Test and Training Range in 2004. Scientists hoped the data would help them improve their theories concerning the origin of the Sun and its planets. NASA deemed one solar wind sample return sufficient for the agency's purposes. If successful, the mission would provide "a reservoir of solar matter for 21st Century science."

[46] 45 SW History, CY 2001, Vol I, pp 103, 104.

[47] Fact Sheet, NASA, "Discovery Mission: GENESIS," updated 1 Apr 2008.

observatories. They were positioned so that one observatory led Earth in its orbit around the Sun while the other observatory trailed it.[48]

NASA's Goddard Spaced Flight Center managed the STEREO mission, enabling NASA to gather data from the observatories, determine spacecraft orbits, and coordinate the scientific results of the mission. The Applied Physics Laboratory designed and built the spacecraft, and the agency also maintained command and control of both STEREO observatories throughout the mission. Each observatory carried a cluster of telescopes, booms and sensors (16 instruments in all). Johns Hopkins University's Applied Physics Laboratory designed each observatory's magnetometers and plasma sensors to: 1) sample outward-bound solar flares (a.k.a. coronal mass ejections) and 2) link the rise of fall of those plasmas and magnetic fields "unambiguously to their origins on the Sun." When the data from the observatories were combined with observations from ground and low-Earth-orbiting sensors, NASA expected to have a three-dimensional picture of earthward-bound solar flares.[49]

The payload's two observatories were launched in a stacked configuration, and they used on-board maneuvering systems to achieve their proper orbital positions after they reached phasing orbit. The launch was successful, and the Applied Physics Laboratory spent the next two weeks checking the observatories' systems to make sure they were working properly. About three months after lift-off, the observatories took up their positions ahead of and behind Earth to continue their two-year-long mission to study solar flares. Officials subsequently extended STEREO's operations into 2009. On 5 May 2009, for example, one of the STEREO observatories detected a Type II radio burst and a coronal mass ejection from the far side of the Sun.[50]

The sixth DELTA II solar mission was nicknamed THEMIS, and it was launched from Pad 17B on 17 February 2007. It featured five identical satellite probes, all of which were placed in orbit approximately one hour and 13 minutes after lift-off. NASA sponsored the Time History of Events and Macroscale Interactions during Substorms (THEMIS) project to explain what triggers the electrically explosive solar wind substorms more commonly known as "the Northern Lights." By deploying five identical probes [51] to magnetically map North America every four days for about 15 hours at a time, scientists hoped to unlock the mysteries of when, where and why solar wind energy was stored in Earth's magnetosphere.[52] NASA funded THEMIS, and the

---

[48] 45 SW History, CY 2006, Vol I, p 80.

[49] 45 SW History, CY 2006, Vol I, pp 80, 81.

[50] 45 SW History, CY 2006, Vol I, pp 80, 83; Fact Sheet, NASA, "STEREO Spacecraft and Instruments," nasa.gov website, 30 Nov 2006; News Release, NASA, "NASA's First 30D Solar Imaging Mission Soars into Space," 25 Oct 2006; NASA, "STEREO Spies First Major Activity of Solar Cycle 24," 15 May 2009.

[51] Each probe was made up of a probe bus and an instrument suite. The probe bus' subsystems included thermal, power, communications, command and control, and guidance and navigation modules. The instrument suite was designed to measure direct current and alternating current, electrical and magnetic fields, and electron and ion distributions and strengths.

[52] As part of the project, technicians at four ground stations in Alaska and 16 ground stations in Canada intended to document the auroras and electromagnetic currents. Once the satellites were checked out and operational on-orbit, approximately 30 substorms would be documented over the next two years.

Goddard Space Flight Center's Explorer Program Office managed the mission. United Launch Alliance (ULA) built the DELTA II launch vehicle and provided the launch service.[53]

Engineers and technicians erected the launch vehicle on Pad 17B between 6 January and 20 January 2007. The THEMIS payload was mated to the DELTA II on 3 February 2007. High upper level winds stymied attempts to launch the vehicle on 16 February, so officials rescheduled the flight for the 17th. The countdown went well, and the DELTA II lifted off the pad at 2301:00.384Z. The five probes separated from the launch vehicle as planned, and mission operators at the University of California (Berkeley) confirmed nominal status via communications with all five spacecraft about an hour later (e.g., 8:07 p.m., Eastern Standard Time). The project hit full stride about a year later when the springtime aurora season opened in 2008. The THEMIS probes had observed one very large geomagnetic storm by that time, and they were well on their way to identifying and tracking the "ropes" of solar wind particles affecting the Earth's magnetosphere.[54]

Asteroid Probes and Comet Encounters

Six DELTA II rendezvous missions with asteroids and comets were launched from Complex 17 between 17 February 1996 and 28 September 2007. The first – nicknamed NEAR, for Near Earth Asteroid Rendezvous – was the first in NASA's 'Discovery' series of flights designed to explore the solar system on a 'shoestring' budget.[55] The entire cost of each Discovery mission – including design, development, the launch vehicle, the payload, launch, flight operations and data analysis – had to be less than $299 million. As the name suggests, the mission was designed to place the NEAR payload in a hyperbolic orbit so it could rendezvous with an asteroid – in this instance, asteroid 433 Eros. The John Hopkins University Applied Physics Laboratory built the spacecraft and managed the mission for NASA.[56]

Technicians and engineers erected the DELTA II's on Pad 17B on between 19 January and 30 January 1996. The third stage and spacecraft assembly were mated to the launch vehicle on 8 February 1996. Authorities scrubbed the first launch attempt on 16 February after a range computer processor failed during the countdown. No show-stoppers surfaced the following day, and the DELTA II lifted off the launch pad at 2043:27.050Z on 17 February 1996.[57]

Over the next four years, the NEAR spacecraft captured images of Earth, the Moon, the Comet Hyakutake, and at least two star clusters (e.g., the Beehive and Pleiades) as it continued

---

[53] 45 SW History, CY 2007, Vol I, pp 92, 93.

[54] 45 SW History, CY 2007, Vol I, p 94; Tony Phillips, GSFC, "Spring is Aurora Season," 4 Mar 2008.

[55] Other missions in the Discovery series included the MARS PATHFINDER, LUNAR PROSPECTOR, STARDUST, GENESIS, CONTOUR, DEEP IMPACT and DAWN missions. In keeping with NASA Director Dan Goldin's philosophy of "faster, better, cheaper" missions, most of the Discovery missions yielded impressive results. Nevertheless, the Mars Climate Orbiter and the Mars Polar Lander mission failures in 1999 drew considerable criticism from the media and other sources who questioned the validity of Goldin's philosophy when it was applied to rocket science.

[56] Fact Sheet, "NASA's Discovery Missions," *Space Today Online*, undated; Fact Sheet, Goddard Space Flight Center, "NEAR Shoemaker, Near Earth Asteroid Rendezvous," undated; New Release, John Hopkins University Applied Physics Lab, "NEAR Shoemaker Makes Historic Touchdown on Asteroid Eros," 12 Feb 2001.

[57] 45 SW History, CY 1996, Vol I, p 92.

on its way to 433 Eros. The spacecraft approached asteroid 253 Mathilde in 1997, and it made its rendezvous with 433 Eros on 14 February 2000. Approaching to within 22 miles of the 21-mile-long asteroid, the NEAR spacecraft used spectrometers and imagers to analyze Eros' composition and capture its principal features. The spacecraft continued to collect and relay data to Earth for almost a full year before controllers commanded the probe to touch down on the surface of 433 Eros on 12 February 2001. The spacecraft collected more than 60 high-resolution images just before it landed on the asteroid, and the spacecraft's beacon continued sending signals until 7:00 p.m. Eastern Standard Time on 28 February 2001.[58]

The next DELTA II asteroid rendezvous mission was launched on 24 October 1998. Its principal goal was to validate the DEEP SPACE 1 spacecraft's celestial navigation system and its remarkable new ion propulsion technology. Along the way, controllers intended to direct the spacecraft to encounter asteroid 1992 KD, the planet Mars, and the comet West-Kohouteck-Ikemura. DEEP SPACE 1 was the first of NASA's new millennium missions, and it heralded NASA's greater emphasis on the development of new technologies to make space exploration more affordable in the future. A microsatellite designed by students from the University of Alabama (Huntsville campus) was carried as a secondary payload.[59] NASA selected a DELTA II Model 7326 vehicle equipped with three GEMs and a Star 37 Upper Stage Yo-Yo system to accomplish the mission.[60]

Technicians and engineers erected the DELTA between 11 September and 18 September 1998. Following a flight simulation on 6 October, the contractor mated the DEEP SPACE 1 spacecraft to the DELTA II on 12 October 1998. Technicians installed the payload fairing on 16 October, and second stage fueling operations were completed on 21 October 1998. There was one unplanned hold during the countdown on 24 October 1998. It was called for a "user processing anomaly" involving the spacecraft. The countdown continued after that problem cleared, and the DELTA II lifted off Pad 17A at 1208:00.502Z on 24 October 1998. The launch was successful, and the spacecraft separated approximately 47 minutes after lift-off.[61]

The three-year-long mission was highly successful. DEEP SPACE 1's lightweight ion engine operated a total of 16,265 hours during the journey, thus proving the value and reliability of its new technology. The 1,071-pound spacecraft encountered asteroid 1992 KD successfully on 20 July 1999, and it accomplished a „fly-by' of asteroid 9969 Braille on 29 July 1999. In lieu of the encounter with the comet West-Kohoutek-Ikemura, the mission was extended to include a close encounter with the comet Borelly on 21 September 2001. DEEP SPACE 1 passed within 1,400 miles of the comet, capturing black and white images of it along with infrared spectrometer data, ion and electron data, and measurements of the magnetic field and plasma

---

[58] Fact Sheet, "NASA's Discovery Missions," *Space Today Online,* undated; Fact Sheet, Goddard Space Flight Center, "NEAR Shoemaker, Near Earth Asteroid Rendezvous," undated; New Release, John Hopkins University Applied Physics Lab, "NEAR Shoemaker Makes Historic Touchdown on Asteroid Eros," 12 Feb 2001.

[59] The payload — SEDSAT (Students for the Exploration and Development of Space Satellite) — was mounted on the DELTA II's second stage. SEDSAT was a six-sided package measuring 13.65 x 13.65 x 12.00 inches. It was designed to collect and distribute data on lightning and other phenomena subject to multi-spectral measurement.

[60] 45 SW History, CY 1998, Vol I, p 93.

[61] 45 SW History, CY 1998, Vol I, pp 93, 94.

waves surrounding the comet. The mission ended officially when the DEEP SPACE 1 spacecraft was powered down on 18 December 2001. It remains in orbit around the Sun.[62]

The first of three DELTA II comet rendezvous missions was launched in February 1999. The object was to send the STARDUST spacecraft on a nearly seven-year-long journey to collect interstellar particles and return them to Earth.[63] Boeing selected a DELTA II Model 7420 vehicle equipped with four GEMs for the mission. According to the flight scenario the launch could occur during only one instantaneous launch window available daily between 6 February and 26 February 1999. The launch actually took place on 7 February 1999. Technicians and engineers erected the DELTA II on Pad 17A between 5 January and 15 January 1999. The contractor mated the spacecraft to the launch vehicle on 28 January 1999. Technicians installed the payload fairing on 2 February, and they loaded propellant into the second stage on 4 February 1999.[64]

The vehicle and its payload were launched at 2104:15.238Z on a flight azimuth of 95 degrees. The first stage and second stage burns were completed during the first 10 minutes of the flight, and the vehicle entered a coasting phase shortly thereafter. Another second stage burn, second/third stage separation, and third stage burn sent the vehicle over the southern end of Africa. The spacecraft separated from the vehicle about 27 minutes after lift-off.[65]

As the STARDUST spacecraft continued on its way through the solar system in the spring of 2000, it collected interstellar dust. It also completed a successful „fly-by' past Earth on 15 January 2001. The spacecraft approached comet Wild-2 on 2 January 2004, and it collected some of the comet's dust as planned. At the end of its 3.2 billion-mile round trip, the spacecraft ejected a return sample capsule that landed in the Utah desert on 15 January 2006. Based on the successful outcome of the mission, NASA officials decided to extend the flight by sending the STARDUST spacecraft on a new mission to rendezvous with comet Tempel 1 in February 2011. The new mission was called Stardust-NEXT, though the spacecraft retained its original name.[66]

The next DELTA II comet encounter mission was the <u>Comet Nucleus Tour</u> (CONTOUR). It was launched from Pad 17A at 0647:41.366Z on 3 July 2002. The object was to place the CONTOUR spacecraft in an orbit that would eventually allow it to encounter at least two of Jupiter's family of comets: 1) Encke (fly-by scheduled for August 2003), and 2) Schwassman-Wachmann-3 (fly-by scheduled for June 2006). The goal was to improve man's knowledge of comet nuclei and their evolution.[67] NASA funded the CONTOUR as a relatively

---

[62] News Release, NASA Jet Propulsion Laboratory, "Deep Space 1 Mission Status," 22 Sep 2001; Fact Sheet, "NASA's first New Millennium mission," *Space Today Online*, 2004.

[63] STARDUST was the first U.S. mission dedicated exclusively to approaching a comet outside the orbit of the Moon. The spacecraft's primary objective was to get close enough to comet Wild-2 to collect some of its dust in the one- to 100-micron range. Secondary objectives included the collection of interstellar particles, imaging the comet's nucleus and coma, and analysis of the particles as STARDUST completed its „fly-through' of the coma. The basic mission ended when the sample return capsule landed at the Utah Test and Training Range in mid-January 2006.

[64] 45 SW History, CY 1999, Vol I, p 76.

[65] 45 SW History, CY 1999, Vol I, pp 76, 77.

[66] Fact Sheet, NASA, "Stardust-NEXT, Exploring Comet Tempel 1," undated version, 9 Jan 2009.

[67] The 2,138-pound spacecraft was equipped with a suite of four major sensors to gather data. The Neutral Gas and Ion Mass Spectrometer (NGIMS) was designed to measure a comet's neutral gas and ambient ions in the coma. The Comet Remote Imaging Spectrograph (CRISP) was supposed to track the comet's nucleus and take high-resolution

low-cost mission under the Discovery program, and the total cost of the spacecraft and its DELTA II Model 7425 launch vehicle was "only" $159 million. Johns Hopkins University's Applied Physics Laboratory (APL) managed the project for NASA, having employed the team that carried out the Near Earth Asteroid Rendezvous (NEAR) mission back in February 1996. Boeing provided the DELTA II, integrated the payload, and launched the mission.[68]

Engineers and technicians erected the DELTA II between 28 May and 1 June 2002. Officials noted abnormal test results for the CONTOUR's second stage pitch actuator during pre-mission testing on 11 June, so technicians replaced the component on that date. The new actuator tested with good results during the simulation flight on 13 June 2002. Engineers mated the CONTOUR spacecraft to the launch vehicle on 19 June, and technicians completed the second stage propellant load on 30 June 2002. Officials certified the vehicle for flight, and preparations continued for the launch on 3 July 2002.[69]

The launch was successful, but the mission ultimately failed. The CONTOUR entered a highly elliptical Earth orbit as planned on 3 July, but Mission Control could not raise the spacecraft after the probe's onboard propulsion system fired on 15 August 2002. NASA asked U.S. Space Command to help locate the missing spacecraft while the APL control team repeatedly tried to establish contact with CONTOUR, but to no avail. Astronomers in Arizona, California, Hawaii and elsewhere managed to photograph at least three pieces of CONTOUR in orbit around the Sun a few days later. The photos verified CONTOUR's Star 30BP solid rocket motor fired as planned, but it apparently malfunctioned about two seconds before the end of its 50-second-long burn. Officials made their final effort to contact CONTOUR on 20 December 2002, but they were unsuccessful. The spacecraft was written off as lost on that date.[70]

The third DELTA II comet mission – DEEP IMPACT – was launched in mid-January 2005.[71] It featured two purpose-built spacecraft destined to complete a six-month-long, 268-million-mile journey to encounter comet 9P/Tempel 1.[72] The larger of the two spacecraft

images. The Comet and Impact Dust Analyzer (CIDA) could measure elemental and chemical composition of dust particles. The Contour Forward Imager (CFI) was built to locate the comet and take color pictures of the nucleus.

[68] 45 SW History, CY 2002, Vol I, p 101.

[69] 45 SW History, CY 2002, Vol I, p 102.

[70] "Boeing Delta Rockets Contour Toward Comets," *Aviation Week & Space Technology*, 8 Jul 2002; "NASA loses contact with craft," *Florida Today*, 16 Aug 2002; "Comet Probe Feared Lost After Maneuver," *Aviation Week & Space Technology*, 19 Aug 2002; "Contour spacecraft seen in sun's orbit," *Florida Today*, 21 Aug 2002; "Hopes Are Fading Fast For Lost Comet Probe," *Aviation Week & Space Technology*, 26 Aug 2002; "Comet craft likely lost; replacement proposed," *Florida Today*, 27 Aug 2002; Press Release, NASA JPL, "CONTOUR Lost," 20 Dec 2002.

[71] Officials planned to launch DEEP IMPACT as early as 30 December 2004, but NASA concluded in late November 2004 that the spacecraft would not be ready in time. Therefore the agency asked for a new launch date of 8 January 2005. Technicians and engineers raised the booster on Pad 17B on 22 November, and they mated the GEMs to the launch vehicle on 1 December 2004. On 10 December, some expert observers noticed DEEP IMPACT's interstage aft structural bracket had not been heat-treated. Engineers erected a new, properly heat-treated interstage on 17 December 2004. Range officials approved a new launch date of 12 January 2005 on 21 December, and launch authorities completed the Simulation Flight on 28 December 2004. Engineers mated the payload to the DELTA II on 3 January, and technicians installed the payload fairing on 7 January 2005.

[72] Ernst Leberecht Tempel discovered comet 9P/Tempel 1 on 3 April 1867. The comet is in an elliptic orbit between Mars and Jupiter, and it circles the Sun every 5.5 years. Astronomers believed the comet was low density, but DEEP IMPACT's encounter revealed Tempel 1 had a lot of rock in it as well.

measured 10.8 x 5.6 x 7.5 feet and weighed 1,325 pounds. The other spacecraft was a 39 x 39-inch cylindrical probe/spacecraft called "the impactor." As its name suggested, the 820-pound impactor was designed to obliterate itself as it created a large crater (reportedly large enough to swallow the Colosseum in Rome) on the face of the four-mile-wide comet.[73] Expelled material from the crater would be observed and recorded by the fly-by spacecraft for data transmission to Earth. Not counting its launch vehicle, the DEEP IMPACT spacecraft and operations associated with the mission cost $267 million. Ball Aerospace & Technologies in Boulder, Colorado, built the spacecraft. NASA's Jet Propulsion Laboratory managed the project, and Dr. Michael A'Hearn led the mission from the University of Maryland.[74]

The countdown went well. There were no unplanned holds, and the DELTA II lifted off Pad 17B at 1847:08.571Z on 12 January 2005. The flight went as planned, and the DEEP IMPACT spacecraft made their rendezvous with comet 9P/Tempel 1 on 4 July 2005. The impactor hit the target area, and the fly-by spacecraft's camera aiming point was within 165 feet of the actual point of impact. The amount of expelled material from the comet was "toward the large end of expectations." The data provided insight into the composition of the comet and, according to NASA, "answered basic questions about the formation of the solar system by offering a better look at the nature and composition of...comets." After passing through the tail of the departing comet, the fly-by spacecraft recorded data from the back of the comet's nucleus and noted any changes in the comet's progress. The major scientific data was transmitted to Earth within 10 hours of the encounter, and supplemental data followed during the next 28 days.[75]

NASA's third DELTA II asteroid rendezvous mission was launched in late September 2007, and it got off to a good start. The eight-year-long DAWN mission was designed to unlock secrets about the early formation of our solar system through a close-up study of two very different celestial bodies – Vesta and Ceres – in the asteroid belt between Mars and Jupiter.[76] Vesta was dry and rocky, displaying evidence of volcanic activity early in its formation. Ceres appeared to have a primitive surface containing water-bearing minerals similar to the large icy moons of the outer solar system. To gather extensive data on both bodies, DAWN was equipped with a specially-designed visible camera, a visible and infrared mapping spectrometer, and a gamma ray and neutron spectrometer.[77]

---

[73] The fly-by spacecraft maneuvered for the rendezvous and released the probe so the impactor could steer itself into the path of the comet. The four-mile-wide comet ran over the 820-pound probe at 22,800 miles per hour, thereby generating sufficient force to create the large crater. The impactor carried a 249-pound copper dead weight as part of its payload to increase the cratering effect.

[74] 45 SW History, CY 2005, Vol I, pp 85, 86; Press Kit, NASA, "Deep Impact Launch," pp 3, 4, 6, Jan 05.

[75] 45 SW History, CY 2005, Vol I, pp 86, 87; Press Kit, NASA, "Deep Impact Launch," pp 3, 4, 6, Jan 05.

[76] Italian astronomer Guiseppe Piazzi discovered Ceres in 1801, and German astronomer Heinrich Wilhelm Olbers located Vesta about six years later. Ceres was the largest object in the asteroid belt, and Vesta (the third largest) was more than 500 kilometers in diameter. Scientists speculated that Ceres had its heaviest minerals at its core and a thick layer of water frozen just below its surface. It was also large enough to have a weak atmosphere and frost-covered polar caps. Vesta appeared to have more in common with planets closer to the Sun, but its formation was literally frozen in time.

[77] 45 SW History, CY 2007, Vol I, p 97.

The countdown on 27 September 2007 went well, despite a 14-minute unscheduled hold for a 'fouled' range. The DELTA II carrying DAWN lifted off Pad 17B at 1134:00.372Z on the 27th, and it injected the spacecraft into the proper Earth-escape trajectory. The DAWN spacecraft completed the first test of its ion propulsion system on 7 October 2007. It was well on its way to Vesta (with a gravity assist from Mars) just a few months later.[78]

Nearly four years after its launch from Cape Canaveral, the DAWN spacecraft is expected to rendezvous with Vesta in September 2011. Once observations and data collection are completed near Vesta in April 2012, DAWN will depart to continue its journey to the "dwarf planet" Ceres. The rendezvous with Ceres in February 2015 could lead to at least five more months of study, in which case the primary mission could be completed in July 2015.[79]

Missions to Mars and Mercury

DELTA IIs supported missions to Mars on eight separate occasions between 7 November 1996 and 5 August 2007. One additional DELTA II boosted a spacecraft on the first leg of its journey to the planet Mercury on 3 August 2004. While two of the Martian missions ended badly, the other flights met or exceeded their sponsors' expectations. Man's continuing fascination with the Red Planet was piqued by each highly publicized Mars mission, and the hard scientific data collected by the successful probes were of lasting value to scientists and educators. The Mercury MESSENGER mission was a significant achievement in its own right, and it will be presented at the end of this chapter.

The DELTA II MARS GLOBAL SURVEYOR (MGS) mission was the first of the „next generation' Martian probes, and it was launched from Pad 17A on 7 November 1996.[80] Lifting off during a pre-planned "instantaneous" launch window, the DELTA II experienced a 30-minute-long coasting period before the third stage and spacecraft were released approximately 43 minutes and 20 seconds into the flight. The third stage Payload Assist Module fired and provided additional thrust for about a minute and a half, and the 2,300-pound MGS separated a little over 50 minutes after lift-off.[81]

The MGS completed its 10-month-long interplanetary cruise at an average speed of 66,000 miles per hour. The spacecraft began its Mars elliptical capture orbit insertion burn at 0117Z on 12 September 1997, and it performed a series of propulsion and aero braking maneuvers to slide itself into a circular polar science mapping orbit of the Red Planet by February 1999. The MGS was designed to collect data on Mars' topography and surface minerals, as well as its gravitational, magnetic and surface remnant fields. Another facet of the

---

[78] 45 SW History, CY 2007, Vol I, p 98.

[79] 45 SW History, CY 2007, Vol I, p 97; Overview, NASA, "Dawn at a Glance," undated; Mark Carreau, "Mission designed to unlock asteroids' secrets," *Houston Chronicle.Com*, 24 Sep 2007.

[80] Technicians and engineers erected the launch vehicle between 20 September and 27 September 1996, and the payload was mated to the launch vehicle on 22 October 1996. Officials scrubbed the first launch attempt on 6 November at 1815Z following an unplanned hold for upper level winds, but the next countdown led to a successful launch at 1700:49.999Z on 7 November 1996. Early indications for a successful mission were reasonably good — one of the spacecraft's solar array panels did not latch properly, but the array was generating "the necessary amounts of power." The MGS was on its way to Mars.

[81] 45 SW History, CY 1996, Vol I, pp 96, 98.

spacecraft's mission was to explore the structure and dynamics of the Martian atmosphere. Data collection in all those areas continued during the spacecraft's repeated orbital passes above the planet's surface for one full Martian year (i.e., 687 days). Once the mapping portion of the mission was completed, NASA planned to use the MGS to relay data from various NASA Mars landers and spacecraft, and officials hoped to continue the operation for three additional years.[82]

The MGS provided remarkable service, and it operated longer in Mars orbit than any other spacecraft in history. After its final communication with Earth on 2 November 2006, the MGS fell silent (an apparent victim of battery failure). The spacecraft studied Mars nearly four times longer than originally planned.[83]

The MARS PATHFINDER (MPF) was launched from Pad 17B aboard a DELTA II on 4 December 1996. The MPF had the distinction of becoming the first American spacecraft to land on Mars since the VIKING missions of the mid-1970s. The launch required precise timing, since there was only one 120-second-long launch window each day to launch the mission between 2 and 28 December 1996. The flight azimuth for each launch opportunity was 95 degrees.[84]

Engineers and technicians erected the DELTA II launch vehicle between 14 October and 22 October 1996. The contractor mated the spacecraft to the DELTA II on 21 November, and technicians installed the payload fairing on 27 November 1996. Authorities scrubbed the first launch attempt on 3 December after a console problem at the pad caused an unscheduled hold at 0701Z. A range support helicopter dropped out during the countdown on the following day, but another range support helicopter covered for it. The countdown continued with no unplanned holds, and the Range supported the DELTA II's successful lift-off at 0658:06.825Z on 4 December 1996.[85]

The flight went as planned. Following a coasting period of nearly one hour, the second stage restarted and fired for approximately 91 seconds. The third stage/payload assembly separated from the second stage about 69 minutes after lift-off. The third stage Payload Assist Module provided thrust for 87 seconds, and the MPF separated from the third stage 75 minutes and 39 seconds into the flight.[86]

Following a direct cruise to Mars, the MPF entered the atmosphere using a VIKING heat shield derivative to slow its descent. A combination of rockets and a parachute aided the MPF's braking maneuver. Once the lander was safely settled on the Red Planet's surface on 4 July 1997, the MPF deployed three solar panels for power. The lander's payload included a camera for data collection and a 35-pound rover for excursions away from the landing area. The environment

[82] "Launch to another world," *45th Space Wing Missileer,* 9 Aug 96; "Mars explorers arrive," *45th Space Wing Missileer,* 23 Aug 1996; "America's Return to Mars," *45th Space Wing Missileer*, 1 Nov 1996; Fact Sheet, Goddard Space Flight Center, "Mars Global Surveyor," undated; Fact Sheet, NASA National Space Science Data Center, "Mars Global Surveyor," ca 2000.

[83] Press Release, NASA Jet Propulsion Laboratory, "Report Reveals Likely Causes of Mars Spacecraft Loss," 13 Apr 2007.

[84] 45 SW History, CY 1996, Vol I, p 98.

[85] 45 SW History, CY 1996, Vol I, p 99.

[86] 45 SW History, CY 1996, Vol I, p 98.

was very harsh, so the lander and rover had operational life expectancies of only 30 days and seven days respectively.[87]

The MPF's performance greatly exceeded its mission planners' expectations. Over a three-month period ending on 27 September 1997, the MPF collected and transmitted 2.3 billion bits of information. The data included weather information, more than a dozen chemical analyses of rocks and soil samples, and more than 17,000 images of Mars. Apart form a few communication 'glitches,' NASA considered the MARS PATHFINDER mission a complete success.[88]

The MARS CLIMATE ORBITER and MARS POLAR LANDER missions were next in line, and they were launched in December 1998 and January 1999 respectively. The missions were complementary, and DELTA II Model 7425 launch vehicles equipped with four GEMs apiece were used for both flights. The MARS CLIMATE ORBITER was designed to monitor the Red Planet's weather and record changes in the Martian surface due to wind and other weather effects. NASA also expected the spacecraft to collect data on temperature changes and dust and moisture content in the Martian atmosphere during a nominal period of one Martian year (i.e., 687 days). The MARS POLAR LANDER's operations were expected to focus on weather conditions near Mars' South Pole. Since the spacecraft was a lander, it was programmed to analyze sample polar deposits for water and carbon dioxide, dig trenches, and collect images of the landing area and the region surrounding it.[89]

Technicians and engineers erected the DELTA II for the MARS CLIMATE ORBITER on Pad 17A between 30 October and 4 November 1998. The spacecraft was mated to the launch vehicle on 30 November 1998. The launch was scheduled for 10 December, but officials scrubbed it on that date for a spacecraft problem. The countdown was recycled and picked up the following day, and the DELTA II lifted off Pad 17A at 1845:51.912Z on 11 December 1998.[90]

In the meantime, engineers and technicians erected the DELTA II for the MARS POLAR LANDER mission on Pad 17B between 28 November and 4 December 1998. The spacecraft was mated to the launch vehicle on 21 December 1998, and the mission was launched at the conclusion of the first countdown at 2021:10.332Z on 3 January 1999.[91]

The launches were successful, but both missions ultimately failed. After losing contact with the MARS CLIMATE ORBITER in September 1999, NASA declared the mission a failure. An official investigation into the mishap later revealed that one team of engineers failed to use metric units in coding a ground software file used for spacecraft trajectory modeling. They used English units which trajectory modelers assumed had been rendered in metric per existing software interface documentation. The spacecraft's Mars insertion trajectory was approximately

---

[87] 45 SW History, CY 1996, Vol I, pp 98, 99; Fact Sheet, NASA Jet Propulsion Laboratory, "Mars Pathfinder Overview," undated; Fact Sheet, MarsNews.com, "Mission Overview: The Design of Pathfinder," ca Jul 2004.

[88] Fact Sheet, NASA Jet Propulsion Laboratory, "Mars Pathfinder Overview," undated.

[89] 45 SW History, CY 1998, Vol I, p 96; 45 SW History, CY 1999, Vol I, pp 74, 75.

[90] 45 SW History, CY 1998, Vol I, pp 96, 97.

[91] 45 SW History, CY 1999, Vol I, p 75.

170 kilometers lower than planned, and the spacecraft either burned up in the Martian atmosphere or re-entered heliocentric space.[92]

Moreover, after losing contact with the MARS CLIMATE ORBITER in September 1999 and ultimately declaring that mission a failure, NASA announced on 6 December 1999 that it could not raise the MARS POLAR LANDER after that spacecraft landed on Mars.[93] Since the MARS POLAR LANDER was not instrumented to provide telemetry as it entered the Martian atmosphere, officials had virtually no data to explain the failure. They speculated that the spacecraft crashed in a canyon in the polar region. The mission was written off as a failure.[94]

Undaunted, NASA launched the MARS ODYSSEY mission from Pad 17A in April 2001, and it proved to be all its sponsors hoped it would be. In this instance, a DELTA Model 7925 equipped with nine GEMs was selected to inject the spacecraft on its flight path to Mars.[95] Timing was crucial, and the launch could occur only during one of two "dual daily" opportunities available between 7 and 30 April 2001.[96] Following lift-off on 7 April 2001, first stage and second stage burns were completed approximately 10 minutes into the flight. Following a coasting period, the second stage restarted and shut down for the second and final time about a minute later. The third stage ignited at T plus 24 minutes and 45 seconds. It provided additional thrust for about a minute and a half before shutting down. The spacecraft separated from the vehicle approximately 31 minutes after lift-off to continue its journey to the Red Planet.[97]

The mission marked a „return to Mars' opportunity for NASA's Langley Research Center,[98] and the flight was flawless. The 1,600-pound spacecraft entered an 18.5-hour capture orbit near Mars on 24 October 2001. The spacecraft used its own 156-pound-thrust engine to maneuver itself into a 15-to-25-hour capture orbit. Aero braking was used thereafter to bring the spacecraft down to a circular, two-hour science orbit.[99] The MARS ODYSSEY spacecraft

---

[92] Mars Climate Orbiter Mishap Investigation Board Phase I Report, 10 Nov 1999, pp 6, 7; "Loss of lander would prompt review," *Florida Today,* 7 Dec 1999, pp 1A, 2A; "MCO Board Probes Deeper Into Flaws," *Aviation Week & Space Technology,* 20 Mar 2000; Discussion, M. Cleary, 45 SW/HO, with Mr. Frank Mann, CSR Plans, 11 Jan 1999.

[93] NASA attempted to reestablish contact with the MARS POLAR LANDER for three days before making the announcement on December 6th.

[94] "Loss of lander would prompt review," *Florida Today,* 7 Dec 1999; Discussion, M. Cleary, 45 SW/HO, with Mr. Frank Mann, CSR Plans, 11 Jan 1999.

[95] Technicians and engineers erected the DELTA II between 26 February and 3 March 2001. The spacecraft was mated to the launch vehicle on 27 March 2001.

[96] The flight azimuth for a launch through either launch window was 65 degrees, but the first window required an inclination of 52 degrees and the second window had an inclination of 49 degrees. A "dogleg" turn occurred shortly after launch to place the vehicle in the proper flight azimuth. The lift-off was recorded at 1502:21.860Z during the first launch opportunity on 7 April 2001.

[97] 45 SW History, CY 2001, Vol I, pp 97, 98.

[98] Langley led Project VIKING in the 1970s, and NASA officials increased the Research Center's participation in the MARS ODYSSEY mission after the loss of the MARS CLIMATE ORBITER and MARS POLAR LANDER spacecraft in 1999.

[99] NASA credited the success to an additional investment of $20 million in the MARS ODYSSEY mission in 2000. The spacecraft's inclination was within .04 degree of its target, and only one 20-minute, 19-second firing of the

maintained that orbit at a mean altitude of 400 kilometers to carry out its primary science mission for the next two and one-half years. During that time, the spacecraft gathered data on Mars' geology and near-space environment via three instruments: 1) the Thermal Emission Imaging System, 2) the Gamma Ray Spectrometer, and 3) the Mars Radiation Environment Experiment. Together with its earlier MARS GLOBAL SURVEYOR mission, NASA learned much about Mars, its mineral resources, and any radiation-related dangers that could affect manned missions to the planet.[100]

MARS ODYSSEY completed its primary mission by September 2004. It continued on an extended mission thereafter. It also served as a communications relay for the Mars Exploration Rovers that landed on the surface of the Red Planet in January 2004.[101]

The MARS EXPLORATION ROVER missions (MER-A and MER-B) proved to be another bright spot in Complex 17's history. The two payloads were virtually identical, and they were launched from Pads 17A and 17B on 10 June and 8 July 2003 respectively.[102] The first spacecraft was carried into space on a DELTA II Model 7925 launch vehicle, but the second payload was launched on a reconfigured DELTA II 7925 Heavy originally erected for the SIRTF mission (see page 39).[103] MER-A separated from its launch vehicle approximately 37 minutes and 14 seconds after lift-off. MER-B – owing to the different launch vehicle and the length of the latter's staging events – separated from the DELTA II Heavy's third stage approximately 80 minutes after lift-off.[104]

The MER-A spacecraft and its twin MER-B were touted as "the most complex spacecraft to come out of the Jet Propulsion Laboratory (JPL) since Cassini was launched to Saturn in 1997." Each rover weighed approximately 400 pounds and carried five main science instruments as well as a rock grinder.[105] Each MER payload consisted of a rover, a lander, an 8.7-foot-

---

spacecraft's onboard engine was required before the start of aero braking on 26 October 2001. Aero braking ended in January 2002, and the probe's mapping mission got underway the following month.

[100] "Odyssey Set To Orbit Mars For Surface Study," *Aviation Week & Space Technology*, 15 Oct 2001; "U.S. Poised For Return to Mars," *Aviation Week & Space Technology*, 2 Apr 2001; "Mars probe launched from Cape," *45th Space Wing Missileer*, 13 Apr 2001; "Odyssey Upgrades Pay Off in 'Bull's-Eye'," *Aviation Week & Space Technology*, 29 Oct 2001; "NASA Begins Odyssey Around The Red Planet," *Space News*, 29 Oct 2001.

[101] Fact Sheet, NASA/JPL, "2001 Mars Odyssey Mission Summary," updated 24 Apr 2007.

[102] According to the flight scenarios, each launch could occur during one of two instantaneous launch windows on a particular launch date. The DELTA II would fly on an azimuth of either 93 or 99 degrees, depending on the window selected. MER-A was launched from Pad 17A at 1758:46.773Z on 10 June 2003. MER-B was launched from Pad 17B at 0318:15.170Z on 8 July 2003.

[103] The booster selected for the MER-B mission was originally erected on Pad 17B in March 2003 to carry NASA's Space Infrared Telescope (SIRTF) into space. However, a GEM de-bonding issue surfaced after the core vehicle was erected, and that problem prompted officials to delay the SIRTF mission until August 2003. In the meantime, officials mated the MER-B on the core vehicle, which remained standing on Pad 17B. A more powerful DELTA II Heavy was needed for the MER-B mission because the payload was launched later in the six-week window of opportunity NASA had selected for the MER-A and MER-B missions.

[104] 45 SW History, CY 2003, Vol I, pp 103, 107, 108, 111; "Red Rover, Red Rover," *Aviation Week & Space Technology*, 26 May 2003; "Fast Action for Rover," *Aviation Week & Space Technology*, 14 Jul 2003.

[105] The main instruments were: 1) a microscopic imager, 2) an alpha particle X-ray spectrometer, 3) a Mossbauer spectrometer, 4) a panoramic stereo camera, and 5) a mini-thermal emission spectrometer.

diameter cruise stage, and an aero shell to carry the lander and rover. Fully fueled, the cruise assembly weighed 2,343 pounds.[106]

Together, the basic MER-A and MER-B missions cost approximately $804 million. The two DELTA II launch vehicles carrying the payloads cost about $60 million apiece. The two payloads arrived and orbited Mars in early January and late January 2004 respectively. Following separate landings on Mars on 4 January and 25 January 2004, the very slow-moving landers[107] assessed Mars' geological features, searched for evidence of water and ancient life, and validated long-distance "rover technologies" and scientific operations of benefit to future Mars explorers. Both rovers completed their three-month-long missions in April 2004, and NASA officials approved a second extension in the mission in late September 2004. Remarkably, both landers were still operating in 2009 following multiple extensions of the twin projects.[108]

The latest of the DELTA II Mars missions was PHOENIX. It was the first of NASA's Scout program missions, which were designed as a competitively priced (and relatively inexpensive) space probe effort. PHOENIX employed a 'leftover' lander from the MARS SURVEYOR project, which officials cancelled in 2001 following the loss of the MARS POLAR LANDER in 1999.[109]

In early March 2007, officials approved the PHOENIX flight for 3 August 2007. A DELTA II Model 7925 was chosen as the launch vehicle, and the spacecraft was launched from Pad 17A on a Type II trajectory toward Mars. Engineers and technicians raised the DELTA II between 18 June and 23 June 2007. Government officials and their contractors completed a simulation flight successfully on 17 July, and engineers mated the spacecraft to the launch vehicle on 23 July 2007. Following the payload fairing installation on 27 July, technicians completed the second stage propellant load on 1 August 2007. Weather delayed the fuel load by one day, so officials rescheduled the lift-off for 4 August 2007.[110]

There were no unplanned holds during the countdown, and the DELTA II lifted off Pad 17A at 0926:34.596Z on 4 August 2007. United Launch Alliance noted that the launch was the 325th flight in the DELTA's long history at Cape Canaveral and Vandenberg AFB. The spacecraft

---

[106] 45 SW History, CY 2003, Vol I, p 103.

[107] Officials hoped each rover might cover one full kilometer of terrain over a period of three months. MER-A's *Spirit* lander drove 3.6 kilometers by September 2004 and MER-B's *Opportunity* lander drove 1.6 kilometers during the same period. Over the next several years, both landers exceeded NASA's expectations by a wide margin.

[108] 45 SW History, CY 2003, Vol I, pp 103, 104; "Red Rover, Red Rover", *Aviation Week & Space Technology*, 26 May 2003; "Batteries Included," *Aviation Week & Space Technology*, 26 May 2003; "Wing aims rocket for Mars," *45th Space Wing Missileer*, 6 Jun 2003; "Mars rovers' plutonium is no threat, NASA says," *Florida Today*, 6 Jun 2000; NASA Facts, "Mars Exploration Rover," pp 1, 2, 3, 4, Oct 2004; JPL, "Spirit Update Archive," 4 Mar 2009 through 20 May 2009; JPL, "Opportunity Update Archive," 4 Mar 2009 through 20 May 2009.

[109] 45 SW History, CY 2007, Vol I, p 95; Overview, NASA, "Mission to Mars, Phoenix Spacecraft," undated; Fact Sheet, NASA, "Mars Phoenix Lander," undated.

[110] 45 SW History, CY 2007, Vol I, p 96.

separated cleanly from its launch vehicle, and PHOENIX was well on its way to Mars a few days later.[111]

After landing on the Red Planet's northern polar region on 25 May 2008, PHOENIX's suite of sensors, robotic arm, portable laboratory and "oven" were used to process water samples retrieved as deep as 20 inches into the planet's icy surface. PHOENIX contained a high-resolution stereo camera to gather data on the Martian atmosphere up to an altitude of 12 miles. It also gathered data on the landing site's geology, including identification of various minerals using multi-spectral sensors.[112]

The mission ended officially in November 2008, but the PHOENIX science team in Tucson, Arizona, continued to analyze PHOENIX's data more than a year after the landing. Researchers hoped to contact the lander successfully in October 2009, but, as of this writing, a response is not anticipated. The mission was, nevertheless, an unqualified success.[113]

Finally we come to the MESSENGER mission, which was launched from Cape Canaveral on a very lengthy and complicated journey to orbit the planet Mercury.[114] The nearly five-billion-mile flight to Mercury entailed an initial Earth fly-by in August 2005, two fly-bys of Venus in October 2006 and June 2007, and three fly-bys of Mercury in January and October 2008 and September 2009.[115] The fly-bys of Mercury were designed to slow down MESSENGER so the spacecraft could enter a highly elliptical 124 x 9,420-mile polar orbit in March 2011. The mission included 412-foot resolution imagery of the entire planet and some 16-foot resolution imagery of Mercury's polar areas. Officials hoped to lower MESSENGER's orbit toward the end of its mission, gathering high-resolution imagery of Mercury before the spacecraft impacted the planet nearly a decade after its launch from Cape Canaveral.[116]

---

[111] 45 SW History, CY 2007, Vol I, p 96; "Robot probe headed for Red Planet," *Air Force Times.Com*, 7 Aug 2007; News Release, NASA, "Phoenix Heads for Mars, Spacecraft Healthy," undated.

[112] 45 SW History, CY 2007, Vol I, p 95.

[113] Overview, NASA, "Mission to Mars, Phoenix Spacecraft," undated; Fact Sheet, NASA, "Mars Phoenix Lander," undated; University of Arizona Lunar and Planetary Laboratory, "Phoenix Team Still at Work as Anniversary Approaches," 20 May 2009.

[114] The project's long title was Mercury Surface, Space Environment, Geochemistry and Ranging. MESSENGER was a NASA/Johns Hopkins Applied Physics Laboratory spacecraft designed to collect data on how Mercury was formed. Eight hundred people representing 50 agencies and companies in seven countries contributed to the development of the $426,000,000 mission. Unlike MARINER 10, which accomplished three Mercury fly-bys in the 1970s, MESSENGER was expected to become the first spacecraft in history to orbit Mercury. MESSENGER weighed 2,425 pounds including the 1,323 pounds of hydrazine and nitrogen tetroxide propellant it needed to accomplish its mission. In addition to the spacecraft's substantial fuel load, the DELTA II 7925H-9.5 vehicle used to boost the spacecraft featured ATK solid rocket motors that were considerably larger than standard DELTA II solid rocket motors.

[115] In early April 2004, NASA officials decided to delay the MESSENGER launch from 11 May 2004 to not earlier than 30 July 2004 to allow more time for final assembly, checkout and testing of the MESSENGER spacecraft's software. MESSENGER's original flight plan was changed from three Venus and two Mercury fly-bys to *two* Venus and *three* Mercury fly-bys, plus an Earth fly-by for a "gravity assist" to compensate for Mercury's change in location during the delay. The change in the flight plan also meant that MESSENGER would arrive on-orbit around Mercury in March 2011 instead of July 2009.

[116] 45 SW History, CY 2004, Vol I, pp 97, 98; "Hot Shot," *Aviation Week & Space Technology*, 26 July 2004; "Feeling the Heat," *Aviation Week & Space Technology*, 8 Aug 2004.

Engineers and technicians erected the DELTA II booster on Pad 17B by the end of June 2004, and the spacecraft was mated to the launch vehicle on 21 July 2004. Officials completed the simulation flight on 15 July, but the launch pad's main crane electrical control box was damaged the following day, and technicians had to isolate and replace a faulty circuit card before the crane went back into operation a few days later. The first countdown on 2 August 2004 went ahead as planned, but bad weather compelled authorities to scrub the launch at 0609Z for a violation of anvil cloud and thick cloud rules. The count was recycled 24 hours, and MESSENGER finally lifted off during a 12-second window at 0615:56.537Z on 3 August 2004. The launch vehicle rolled into a flight azimuth of 107 degrees, and the MESSENGER spacecraft separated from the DELTA II approximately 56 minutes and 43 seconds after lift-off.[117]

By 8 August 2004, MESSENGER was 1.5 million miles away from Earth and "safely en route" to its destination. Controllers at Johns Hopkins' APL Mission Operations Center in Laurel, Maryland, continued to check out the spacecraft via NASA's Deep Space Network antennas at least three times per week during the course of the flight. MESSENGER flew by Mercury in January 2008 as expected, and the probe collected more than 1,200 high-resolution images and color photos of the planet's surface during the second fly-by on 6 October 2008. The spacecraft surveyed 30 percent of the planet that had been hidden from earlier probes, and it discovered a 430-mile diameter impact basin heretofore unknown. Other findings included the first indication of magnesium in Mercury's exosphere, a better understanding of the little planet's internal magnetic field and its interaction with the solar wind, and more precise data related to Mercury's surface geology. The spacecraft also received a critical gravity assist it needed to complete its third fly-by in September 2009, followed by the final leg of its journey to orbit Mercury in 2011.[118]

Kepler Spacecraft 'Planetary Search' Mission

Unlike the previous four categories of missions, the Kepler Spacecraft „planetary search' mission was designed to: 1) survey approximately 100,000 stars in our corner of the Milky Way galaxy, and 2) detect previously undiscovered planets.[119] By various means since 1995, scientists had detected approximately 330 planets orbiting other stars, but most of them were 'gas giants' similar to Jupiter or Neptune. Others were ice giants or super-hot Earths. None of the newly

---

[117] 45 SW History, CY 2004, Vol I, p 98.

[118] "Safety Slip," *Aviation Week & Space Technology*, 26 Jul 2004; "Feeling the Heat," *Aviation Week & Space Technology*, 8 Aug 2004; "Mission to Mercury," *45th Space Wing Missileer*, 6 Aug 2004; News Release 08-275, NASA and Johns Hopkins University Applied Physics Laboratory, "Messenger Spacecraft Reveals More Hidden Territory on Mercury," 29 Oct 2008; News Release 09-092, NASA and Johns Hopkins University Applied Physics Laboratory, "Messenger Spacecraft Reveals a Very Dynamic Planet Mercury," 30 Apr 2009; Messenger Teleconference Introduction, NASA, "MESSENGER Reveals More About The Evolution, Tectonics Of Mercury," 30 Apr 2009.

[119] Utilizing a 0.95-meter aperture photometer with a 12-degree-diameter field of view, the Kepler's array of 42 Charge Couple Devices (CCDs) were designed to detect Earth-sized planets orbiting stars in our region of the Milky Way. From its undisturbed vantage point on-orbit, the Kepler three-axis stabilized spacecraft might discover hundreds of Earth-sized and smaller planets orbiting in or near the "habitable" zone (i.e., where liquid water might form on a planet's surface owing to the latter's composition and distance from its sun) of some stars in the region. Additionally, the Kepler mission was undertaken to explore the structure, diversity, and distribution of planetary systems more generally, perhaps estimating how many planets exist in multiple-star systems, what percentage of Earth-sized or larger planets were in the habitable zone, what types of stars harbored planetary systems, etc.

detected planets were considered hospitable to life. Nevertheless, those discoveries prompted scientists to wonder about the number of rocky, Earth-like planets that might be out there – and how many of them that we might be able to detect. Consequently, NASA sponsored the $591 million Kepler Spacecraft mission as one of the agency's Discovery missions (#10). Over a period of three and one-half years, the large sampling of data gleaned by the Kepler Spacecraft on-orbit could give NASA scientists a much better idea of how rare – or common – Earth-like planets are, at least in our little corner of the Universe.[120]

Ball Aerospace & Technologies Corporation (BATC) built the Kepler Spacecraft bus and photometer. The spacecraft was equipped with four solar panels to provide 1,100 watts of power. The Kepler featured a radiation-shielded PowerPC flight computer, and it was equipped with a Ka-Band communications link to transfer data back to Earth. The photometer was basically a Schmidt telescope design, and the spacecraft served it as a source for power, pointing, and telemetry. The spacecraft was 15.3 feet long, and it weighed 2,320 pounds. The United Launch Alliance DELTA II Model 7925-10L launch vehicle selected for the mission was equipped with nine GEMs and a Star 48B third stage. According to the flight scenario, The DELTA II carrying the Kepler Spacecraft was launched from Pad 17B on a flight azimuth of 93 degrees. Following the first stage and second stage burns, the DELTA II and its payload entered a 100-nautical-mile circular orbit. An additional second stage burn was required before the Star 48B upper stage fired for five and one-half minutes to boost the Kepler Spacecraft into an Earth-trailing heliocentric orbit.[121]

Engineers and technicians erected the DELTA II's first stage on Pad 17B on 21 October 2008, and the GEMs were added to the launch vehicle by 4 November 2008. As processing continued, the second stage was erected on 17 December 2008, and officials completed the Simulation Flight on 29 January 2009. Workers delivered the Kepler Spacecraft to the launch pad from Astrotech in Titusville, Florida, on 19 February, and engineers mated the payload to the DELTA II on 21 February 2009. Technicians installed the payload fairing on 26 February 2009. Analysis of a TAURUS payload fairing failure delayed the second stage propellant loading operation by one day, so the launch date was pushed from 6 March to 7 March 2009.[122]

NASA had only two, three-minute-long 'windows' to launch the mission on the 7th, but there were no unscheduled holds, and the lift-off was recorded at 0349:57.465Z. Following the spectacular night launch (i.e., the launch occurred at 10:49 p.m. Eastern Standard Time), the spacecraft separated from the Star 48B third stage about one hour after lift-off. A tracking station in California confirmed Kepler's solar panels were generating power shortly thereafter. Following 60 days of calibrations and testing, the Kepler Spacecraft could begin hunting for

---

[120] Summary (U), NASA, "Overview of Kepler Mission," *kepler.nasa.gov,* undated; Summary (U), NASA, "Photometer and Spacecraft," *kepler.nasa.gov,* undated; "NASA Prepares Kepler For Launch," *redorbit.com,* 1 Feb 2009; Will Dunham, "NASA spacecraft to seek out Earth-like planets," *reuters.com,* 19 Feb 2009; News Release, NASA, "NASA's Kepler Mission Rockets to Space in search of other Earths," 7 Mar 2009; William Harwood, "Kepler spacecraft leaves Earth to discovery new worlds," *spaceflightnow.com,* 6 Mar 2009.

[121] 45 SW History, CY 2009, Vol I, (Draft) p 76; William Harwood, "Kepler spacecraft leaves Earth to discovery new worlds," *spaceflightnow.com,* 6 Mar 2009; Summary, NASA, "Kepler Mission: Launch Vehicle and Orbit," *kepler.nasa.gov,* undated.

[122] 45 SW History, CY 2009, Vol I, (Draft) p 76; James Dean, "Kepler Reaches Cape Launch Pad," *floridatoday.com,* 19 Feb 2009.

planets as planned. Traveling at five miles per second, Kepler passed the Moon within a couple of days as it continued on its own heliocentric orbit. By March 2010, it would be approximately 10 million miles from Earth. The Kepler Spacecraft mission was off to a good start.[123]

---

[123] 45 SW History, CY 2009, Vol I, (Draft) p 77; Todd Halvorson, "Planet-hunting satellite lifts off tonight," *floridatoday.com,* 6 Mar 2009; James Dean, "Live At The Cape: Kepler Launch A Success," 6 Mar 2009; Sharon Gaudin, "NASA's Kepler spacecraft hurtles past moon's orbit," *computerworld.com,* 9 Mar 2009.

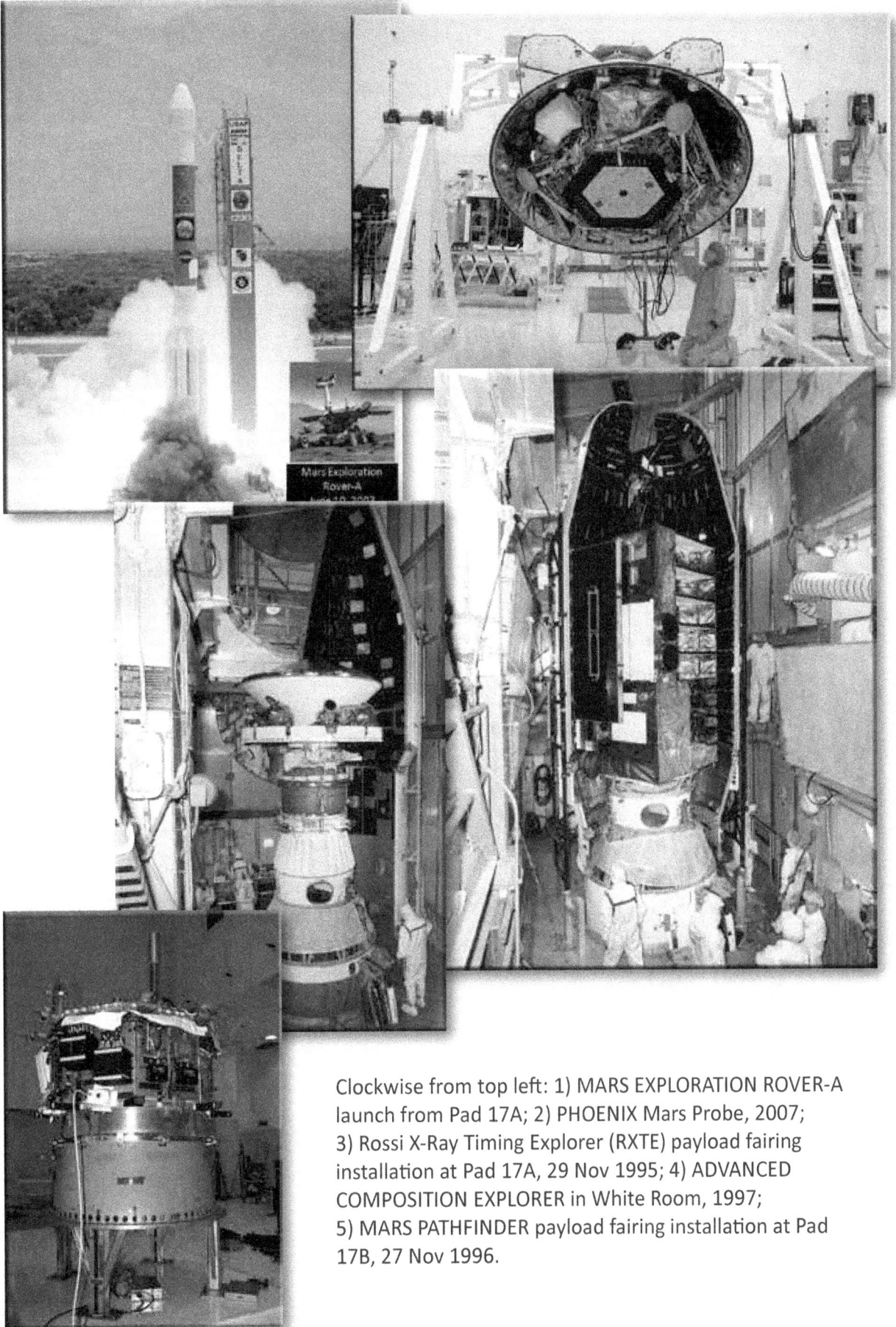

Clockwise from top left: 1) MARS EXPLORATION ROVER-A launch from Pad 17A; 2) PHOENIX Mars Probe, 2007; 3) Rossi X-Ray Timing Explorer (RXTE) payload fairing installation at Pad 17A, 29 Nov 1995; 4) ADVANCED COMPOSITION EXPLORER in White Room, 1997; 5) MARS PATHFINDER payload fairing installation at Pad 17B, 27 Nov 1996.

Clockwise from top left: 1) a DELTA II carrying the DAWN asteroid probe lifts off Pad
17B (note insert); 2) DAWN spacecraft; 3) DELTA III engine in Hangar AO, 7 Dec 1998;
4) DELTA III ORION-3 second stage is erected on Pad 17B, 6 Mar 1999; 5) DELTA III
ORION-3 lift-off, 5 May 1999.

CHAPTER IV

COMMERCIAL MISSIONS

McDonnell Douglas launched a DELTA II commercial communications satellite mission – PALAPA B2R – from Cape Canaveral in mid-April 1990. That flight was the first of 23 DELTA II commercial payload missions to lift off Complex 17 over the next decade. Except for the KOREASAT-1 mission in 1995, all those flights were completely successful. DELTA IIs continued to carry commercial payloads after Boeing bought McDonnell Douglas in 1997, but the parent company also fielded a new and larger launch vehicle, the DELTA III, to break into a heavier commercial launch service market. Three DELTA III missions lifted off Pad 17B between 27 August 1998 and 24 August 2000. Two of the DELTA III flights were unsuccessful, but Boeing clearly recognized that the commercial spacecraft industry was moving toward larger, heavier and more expensive payloads that could not be carried into space on DELTA IIs. Larger launch vehicles – like the company's own DELTA IV and Lockheed Martin's ATLAS V – were needed. Sadly, the DELTA III could not compete successfully in the new market, but the DELTA IV proved to be a worthy successor to the incredibly successful DELTA II.

The missions presented below have been grouped according to their customers. Since many of the spacecraft were virtually identical or fell into series of satellites bought for the same satellite constellation, it makes sense to group the missions in this manner. Appendix C lists them in chronological order if any further sorting is required to make sense of the sequence of events.

PALAPA B2R and PALAPA B4

The PALAPA B2R was one of two payloads that failed to achieve transfer orbit during the Space Shuttle *Challenger's* STS-41B mission in February 1984, but it was recovered along with Western Union's WESTAR IV satellite during the STS-51A mission in November 1984.[1] It was sent back to Hughes' Space and Communications Group where it was refurbished for redeployment under a contract with Indonesia's Perusahaan Umum Telekomunikasi (PERUMTEL). The spacecraft was a Hughes Model HS-376, and it was designed to provide 24 channels of high quality communications to Indonesia and other members in the Association of Southeast Asian Nations (ASEAN). The satellite was approximately 86 inches in diameter, and its length was approximately 112 inches in stowed configuration. With its outer solar panel cylinder extended on-orbit, it was nearly 23 feet long. The spacecraft had a mass of approximately 1,525 pounds on-orbit, but its weight at launch (including hydrazine fuel) was considerably greater. PALAPA B-series spacecraft were expected to operate on-orbit for at least nine years.[2]

The DELTA II Model 6925 carrying the PALAPA B2R lifted off Pad 17B at 2227:59.719Z on 13 April 1990. After rolling into a flight azimuth of 95 degrees, the launch vehicle completed

---

[1] Both satellites were deployed successfully in February, but their Payload Assist Modules (PAMs) cut off early, leaving them in useless orbits ranging from 550 to 575 miles above Earth. They were maneuvered into lower orbits a few days before *Discovery* was launched on 8 November 1984. Mission specialists Joseph P. Allen and Dale A. Gardner retrieved the PALAPA B2R and WESTAR IV satellites successfully on 12 November and 14 November 1984 respectively.

[2] ESMC History, FY 1990, Vol I, p 316; ESMC History, 1 Oct 1982 – 30 Sep 1984, Vol I, p 194; ESMC History, 1 Oct 1984 – 30 Sep 1986, Vol I, pp 237, 238, 240; Fact Sheet, Hughes Space and Communications Group, "PALAPA-B, Indonesia's Second Generation Satellite," Mar 1992.

its initial second stage burn approximately 11 minutes into the flight. Following a coasting phase and a second stage restart, the Star 48B third stage fired for approximately 87 seconds, and the spacecraft separated from the launch vehicle about 26 minutes after lift-off. The PALAPA B2R achieved the proper transfer orbit apogee approximately five and one-half hours later. Unlike the spacecraft's first trip into space, this flight was completely successful. After replacing the PALAPA B-1[3] on station at 108 degrees East longitude, the PALAPA B2R provided commercial communications services until its operations ended in 2000.[4]

A DELTA II Model 7925 carrying another Hughes Model HS 376 satellite for Indonesia was launched on 14 May 1992. Named PALAPA B4, it was virtually identical to the PALAPA B2R, and it was destined to join its sibling on-orbit at 118 degrees East longitude. Only two brief holds for high-level winds and heavy clouds marred the otherwise uneventful countdown, and the DELTA II lifted off Pad 17B at 0040:00.058Z on the 14th. After rolling into a flight azimuth of 97.5 degrees, the launch vehicle injected the spacecraft into a 150 x 19,923 nautical-mile transfer orbit. The mission was successful, and the PALAPA B4 provided commercial communications services to ASEAN member nations through the end of its on-orbit operations in 2005.[5]

The next generation of Indonesian commercial communications satellites included TELKOM-1 and TELKOM-2. They were based on the A2100A Lockheed Martin and Starbus 2 spacecraft series respectively. TELKOM-1 was launched on an ARIANE 4 from Kourou, French Guiana, in August 1999. It replaced the PALAPA B2R on-orbit. TELKOM-2 was launched from Kourou on an ARIANE 5 on 16 November 2005, and it replaced the PALAPA B4 on-orbit.[6]

Britain's BSB -R2 and Telenor's THOR II and THOR III

As noted in Chapter I, Britain's BSB-R1 spacecraft was the first commercial payload ever launched into orbit by a U.S. commercial space launch vehicle. The vehicle selected for that mission was an older-generation DELTA, but the next satellite in the series – BSB-R2 – was launched from Pad 17B in August 1990 on a brand new DELTA II. The BSB-R2 was a Hughes Model HS-376 spacecraft, and it was designed to provide five-channel direct-broadcast television for the United Kingdom. It was much like the PALAPA B2R and PALAPA B4 spacecraft in weight, size, and configuration, and it had an on-orbit life expectancy of 10 years.[7]

The countdown for the BSB-R2 mission on 17 August 1990 was hampered by heavy thunderstorm activity, but the DELTA II lifted off the pad successfully at 0041:59.891Z on 18 August 1990. The vehicle rolled into a flight azimuth of 95 degrees, and the spacecraft was released into a 100 x 20,572 nautical-mile transfer orbit as planned. In an interesting turn of

---

[3] The PALAPA B1 was launched aboard the Space Shuttle *Challenger* in June 1983. The satellite was retired from service in September 1990.

[4] ESMC History, 1 Oct 1982 – 30 Sep 1984, Vol I, pp 184, 185; ESMC History, FY 1990, Vol I, p 316; Tonda Priyanto, for TELKOM, "The Journey of TelKom in Operating Communications Satellites to Serve the Indonesian Archipelago," *Online Journal of Space Communication,* Issue No.8, Fall 2005; Fact Sheet, Hughes Space and Communications Group, "PALAPA-B, Indonesia's Second Generation Satellite," Mar 1992.

[5] 45 SW History, CY 1992, Vol I, p 233; Priyanto, "The Journey of TelKom," Fall 2005; Fact Sheet, Hughes Space and Communications Group, "PALAPA-B, Indonesia's Second Generation Satellite," Mar 1992.

[6] Priyanto, "The Journey of TelKom," Fall 2005; Fact Sheet, Orbital Sciences Corporation, "TELKOM-2," 2009.

[7] ESMC History, CY 1990, Vol I, p 319; News Release, McDonnell Douglas, "McDonnell Douglas Gets Contract to Launch Thor III Satellite on a Delta II rocket," 29 Jul 1997.

events, British Satellite Broadcasting was taken over by Sky Television to become British Sky Broadcasting in the early 1990s, and the latter sold the BSB-R2 spacecraft to a Scandinavian satellite communications firm named Telenor in 1992. The spacecraft was renamed THOR I, and it eventually provided nearly 10 years of continuous service to more than 750,000 customers in Scandinavia. It was taken out of service in late 2001.[8]

Hughes Space and Communications International received contracts to build Telenor's THOR II and THOR III (i.e., two more HS 376 spacecraft) in 1995 and 1997 respectively. McDonnell Douglas launched THOR II successfully on 20 May 1997 following two launch scrubs just days earlier. Two months later, the company announced it had a contract to launch THOR III in July 1998. As events unfolded, Boeing launched THOR III from Pad 17A at 0035:00.571Z on 10 June 1998. The THOR II and THOR III were still in service in 2008, but THOR II was due for immediate replacement, and THOR III might be replaced within a few years.[9]

ASC-2, AURORA II and SATCOM 4

Three American commercial communications spacecraft – ASC-2, AURORA II and SATCOM 4 – were launched on DELTA IIs from Pad 17B between 13 April 1991 and 1 September 1992. The ASC-2 spacecraft was purchased by CONTEL to provide telecommunications services to the continental United States. The satellite carried 18 C-Band channels and six wideband KU-Band channels, and it operated from geosynchronous orbit approximately 22,000 miles above the equator. The AURORA II spacecraft was purchased by Alascom, Inc. to provide commercial telecommunications services from geosynchronous orbit to Alaska on 24 C-Band channels. The SATCOM 4 was owned by GE American Communications, Inc., and it provided 24 channels of cable video services from geosynchronous orbit to all 50 of the United States as well as the Caribbean. ASC-2 and AURORA II were launched on a flight azimuth of 95 degrees, and SATCOM 4 was launched on a flight azimuth of 93 degrees. All three spacecraft had on-orbit life expectancies of 12 years.[10]

All three flights were successful. Officials extended the countdown for the ASC-2 launch for 31 minutes for excessive upper level winds, but the vehicle lifted off without incident at 0009:00.027Z on 13 April 1991. The spacecraft separated from the vehicle about 30 minutes after launch. The AURORA II countdown on 29 May 1991 was delayed for 31 minutes to replace a pad camera, but the DELTA II was launched at 2255:00.037Z on that date. The spacecraft was injected into a 765 x 19,323 nautical-mile transfer orbit by the vehicle's third stage approximately 30 minutes after lift-off. Last but not least, the countdown for the SATCOM 4 mission proceeded smoothly on 31 August 1992, and the DELTA II was launched at 1040:59.083Z on that date. The SATCOM 4 was injected in a highly elliptical transfer orbit after the third stage burned out, approximately 77 minutes after lift-off.[11]

---

[8] ESMC History, CY 1990, Vol I, pp 319, 320; News Release, McDonnell Douglas, "McDonnell Douglas Gets Contract to Launch Thor III Satellite on a Delta II rocket," 29 Jul 1997; Press Release, Telenor, "Telenor's First Satellite, Thor I, at the end of the road," undated.

[9] News Release, McDonnell Douglas, "McDonnell Douglas Gets Contract to Launch Thor III Satellite on a Delta II rocket," 29 Jul 1997; Boeing, "Hughes Builds Two TV Satellites for Norway," Nov 1997; 45 SW History, CY 1997, Vol I, pp 98, 99; ; 45 SW History, CY 1998, Vol I, p 93.

[10] 45 SW History, 1 Oct 1990 – 31 Dec 1991, Vol I, pp 343, 344, 345; 45 SW History, CY 1992, Vol I, p 236.

[11] 45 SW History, 1 Oct 1990 – 31 Dec 1991, Vol I, pp 344, 345; 45 SW History, CY 1992, Vol I, p 236.

INMARSAT -2 F-1 and INMARSAT -2 F-2

Two telecommunications satellites – INMARSAT-2 F-1 and INMARSAT-2 F-2 – were launched from Pad 17B for the International Maritime Satellite Organization on 30 October 1990 and 8 March 1991. The INMARSAT-2 payloads were virtually identical, and their DELTA II Model 6925 launch vehicles rolled into a flight azimuth of 95.85 degrees following lift-off. The staging sequences were timed to ensure that both spacecraft were placed in 100 x 19,750 nautical-mile transfer orbits. British Aerospace Space Systems Ltd built the INMARSAT-2s for approximately $90 million apiece. The satellites were placed on-orbit to join INMARSAT's four-satellite constellation 22,300 miles over the Atlantic Ocean. Each satellite carried four ship-to-shore antennas. Services included telephone, telex, data, and facsimile communications. Each spacecraft had an on-orbit life expectancy of 10 years.[12]

The countdown for the INMARSAT-2 F-1 mission went smoothly with no unscheduled holds, and the DELTA II lifted off at 2316:00.140Z on 30 October 1990. The spacecraft separated from the launch vehicle approximately 44 minutes after lift-off. There was a four-minute extension during the final built-in hold for the INMARSAT-2 F-2 launch, but the DELTA II lifted off successfully at 2302:59.960Z on 8 March 1991. The spacecraft separated from the launch vehicle's third stage about 45 minutes and 31 seconds after lift-off.[13]

DFS KOPERNIKUS 3

A DELTA II carrying the DFS KOPERNIKUS 3 spacecraft lifted off Pad 17B at 0947:00.012Z on 12 October 1992. The payload was one of three German satellites (Deutscher Fernmelde Satellit – DFS) owned by Deutsche Bundepost Telekom, the Germany Post Office's telecommunications division. MBB/ERNO[14] built the spacecraft, which was equipped with 11 transponders to operate in three frequency ranges from geosynchronous orbit. The satellite used a Star 30C solid rocket motor to achieve geosynchronous orbit, and the spacecraft had an on-orbit life expectancy of 10 years.[15]

There were no unscheduled holds during the countdown, and the DELTA II Model 7925 launch vehicle lifted off the pad and rolled into a flight azimuth of 93 degrees. The spacecraft separated from the launch vehicle approximately 80 minutes after lift-off. DFS KOPERNIKUS 3 provided yeoman service over the next decade, but it was retired in 2002. It was subsequently moved a safe distance from functioning satellites and placed in on-orbit storage.[16]

GALAXY I -R and GALAXY IX

The GALAXY I-R and GALAXY IX missions were launched from Pad 17B on 19 February 1994 and 24 May 1996 respectively. Both payloads were Hughes Model HS 376 spacecraft built to join Hughes Communications' constellation of GALAXY satellites.[17]

---

[12] 45 SW History, 1 Oct 1990 – 31 Dec 1991, Vol I, pp 339, 340, 341, 342.

[13] 45 SW History, 1 Oct 1990 – 31 Dec 1991, Vol I, pp 339, 340, 341, 342, 343.

[14] Messerschmitt-Bölkow-Blohm (MBB) acquired Entwicklungsring Nord (ERNO) in 1982 to form MBB/ERNO. Observers agreed that substituting the acronym for the full name made conversations easier, and the short title saved considerable time and expense in lettering on buildings, stationery, equipment, etc,.

[15] 45 SW History, CY 1992, p 237.

[16] 45 SW History, CY 1992, p 237; Peter de Seiding, "FCC Enters Orbital Debris Debate," *Space News*, 28 Jun 2004.

GALAXY I-R was added to the constellation as Hughes' second cable television-dedicated satellite. (It replaced GALAXY I, which was near the end of its service life.) Hughes' Mission Control Center in El Segundo, California, controlled and monitored the spacecraft once it was on-orbit. The GALAXY I-R weighed 3,070 pounds at lift-off and 1,700 pounds in geostationary orbit. The spacecraft was 85 inches in diameter and 24 feet and 8.5 inches high in its extended on-orbit configuration. The satellite's 24 C-Band transponders were designed to provide high quality communications services to the continental United States.[18]

Technicians and engineers erected the GALAXY I-R's launch vehicle between 12 January and 21 January 1994. Engineer added the spacecraft on 29 January, and the first launch attempt got underway on 9 February 1994. Officials scrubbed the first countdown after the engine start command failed to start the DELTA II's first stage engine. The weather was not favorable for the second launch attempt on 18 February, and technicians had to complete GEM ignition and separation connections with the Mobile Service Tower wrapped around the vehicle for wind protection. Officials scrubbed the second countdown at 1715Z, and the countdown was recycled for 19 February 1994. The vehicle was left as it stood, pending the third launch attempt the following day. Officials extended the third countdown 33 minutes for rain showers over the launch pad, but the count proceeded uneventfully thereafter. The DELTA II lifted off the launch pad at 2344:59.741Z on 19 February 1994, and the spacecraft separated from the launch vehicle approximately 81 minutes and 48 seconds into the flight.[19]

The GALAXY IX mission was well underway two years later. Engineers and technicians erected GALAXY IX's launch vehicle between 3 April and 9 April 1996. The contractor mated the spacecraft to the DELTA II on 8 May 1996, and officials completed the Launch Readiness Review on 21 May 1996. The countdown on 23 May was largely uneventful, but there was a 35-minute extension in one of the pre-planned holds due to "time constraints." The countdown resumed, and the DELTA II lifted off Pad 17B at 0109:59.861Z on 24 May 1996. The vehicle rolled into a flight azimuth of 97.5 degrees, and the spacecraft separated from the launch vehicle approximately 80 minutes after lift-off. GALAXY IX became the latest addition to a network of satellites providing communications services to the continental United States, Alaska, Hawaii and Puerto Rico in the 6/4 GHz (C-Band) range.[20]

---

[17] Eight HS 376 spacecraft were purchased for the GALAXY constellation, but only seven reached orbit. The first three satellites were launched on old DELTA rockets in June 1983, September 1983, and September 1984 respectively. Each had an operational life expectancy of nine years on-orbit. The fourth GALAXY spacecraft was launched on an ARIANE rocket from Kourou, French Guiana, in October 1990, and the fifth one was launched on a ATLAS I/CENTAUR vehicle from Pad 36B on 14 March 1992. The sixth GALAXY satellite was designated GALAXY I-R as a replenishment spacecraft, and it was launched on an ATLAS I/CENTAUR from Pad 36B at Cape Canaveral on 22 August 1992. Unfortunately, the CENTAUR upper stage malfunctioned, and the vehicle was destroyed by Mission Flight Control Officers (MFCOs) eight minutes after launch. The new GALAXY I-R that was launched in February 1994 as the seventh HS 376 satellite purchased for the constellation. GALAXY IX was the eighth spacecraft in the series, and the seventh satellite to reach orbit. The GALAXY IX and the surviving GALAXY I-R each had an on-orbit life expectancy of 12 years.

[18] ESMC History, 1 Oct 1982 – 30 Sep 1984, Vol I, pp 212, 213, 217, 220, 221; 45 SW History, CY 1992, Vol I, pp 224, 226; 45 SW History, CY 1994, Vol I, p 104; 45 SW History, CY 1996, Vol I, p 93.

[19] 45 SW History, CY 1994, Vol I, pp 105, 106.

[20] 45 SW History, CY 1996, Vol I, pp 93, 94.

KOREASAT -1 and KOREASAT -2

The KOREASAT-1 and KOREASAT-2 missions were launched from Pad 17B on 5 August 1995 and 14 January 1996 respectively. The object of each flight was to place a KOREASAT spacecraft in a 1,353 x 35,786-kilometer geosynchronous transfer orbit. Lockheed Martin manufactured both satellites in East Windsor, New Jersey. Korea Telecom sponsored the spacecraft to provide direct broadcasting and telecommunications services to customers throughout the Republic of Korea. Each spacecraft was equipped with 12 transponders for video broadcasts, inter-city trucking, multi-speed data transfer, and VSAI services. Three other transponders were included in the satellite for direct television broadcasting. Each satellite had an on-orbit life expectancy of 10 years.[21]

Before it was brought out to the launch pad, the KOREASAT-1 spacecraft was processed and mated to its Star 48B upper stage at Astrotech's facilities in Titusville, Florida. Technicians and engineers erected the DELTA II between 18 May and 26 May 1995. Engineers erected the spacecraft on 24 July, and technicians installed the payload fairing on 28 and 29 July 1995. Though there was a mission rehearsal on 1 August, officials had to batten down the launch site for Hurricane ERIN on 2 August, and technicians had to reconfigure the launch vehicle and site before resuming preparations for the launch. As the situation unfolded, the threatening storm did not impact the Cape, but the launch was delayed from 3 August to 5 August 1995. There were no unplanned holds during the countdown, and the DELTA II lifted off the launch pad at 1110:00.028Z on 5 August 1995.[22]

The flight sequence went very well initially, but one of the DELTA II's three air-lit GEMs failed to separate from the vehicle. The additional weight diminished the vehicle's overall performance, and the payload was placed in a transfer orbit about 3,000 miles lower than planned. KOREASAT-1's on-orbit operational life may have been reduced approximately four years as a result of the flight anomaly, but the spacecraft's onboard propellant reserve was sufficient to raise it to a useful orbit. (The satellite was retired in 2005.) McDonnell Douglas quickly formed an anomaly team to investigate the incident and determine its cause. Planners needed to ensure the launch vehicle's performance was up to specifications before the next DELTA II mission was launched later in the year. The next DELTA II placed the Rossi X-Ray Timing Explorer (RXTE) in the proper transfer orbit on 30 December 1995, and the successful launch restored confidence in the DELTA II stable of launch vehicles.[23]

Korea Telecom proved willing to give the DELTA II another chance, and the company allowed the KOREASAT-2 mission to go forward in January 1996. Engineers and technicians erected the launch vehicle's first and second stages on 13 and 14 November 1995, and they attached the GEMs between 2 December and 8 December 1995. Final GEM alignments were completed on 21 December 1995. Engineers mated the third stage/spacecraft assembly to the launch vehicle on 5 January 1996, and the payload fairing was installed on 9 January 1996. Following the mission rehearsal on 10 January, preparations continued for the launch.[24]

---

[21] 45 SW History, CY 1995, Vol I, p 107; 45 SW History, CY 1996, Vol I, p 90.

[22] 45 SW History, CY 1995, Vol I, p 107.

[23] 45 SW History, CY 1995, Vol I, pp 107, 109, 111; Stephen Clark, "Satellite with double use launched for South Korea," *Spaceflight Now,* 22 Aug 2006.

[24] 45 SW History, CY 1996, Vol I, p 91.

There was one unscheduled hold during the countdown on 14 January for a DELTA II boat tail air temperature out of tolerance indication, but the problem was corrected by readjusting a damper to bring the temperature within specifications. That problem aside, the countdown went smoothly, and the DELTA II lifted off the launch pad at 1110:00.149Z on 14 January 1996. The launch vehicle rolled into a flight azimuth of 93 degrees shortly after launch. The DELTA II's staging sequences were completely normal, and the vehicle's GEMs separated cleanly at the appropriate times. The KOREASAT-2 spacecraft separated from the third stage approximately one hour and 17 minutes after lift-off, and the launch was completely successful. KOREASAT-2 was still providing service 10 years later.[25]

BONUM-1

The object of the DELTA II flight in late November 1998 was to place the BONUM-1 communications satellite in a nominal transfer orbit. Hughes Space and Communications Company built the HS 376 spacecraft to provide high-quality communications to Russia and its neighbors in the 18/12 GHz (Ku-Band) frequency range.[26] Boeing selected a DELTA II 7925 launch vehicle equipped with nine GEMs to accomplish the mission. Technicians and engineers erected the DELTA II on Pad 17B between 19 October and 30 October 1998. Following a flight simulation on 6 November, engineers mated the BONUM-1 spacecraft to the DELTA II on 11 November 1998. Technicians installed the payload fairing on 14 November, and second stage fueling operations were completed on 17 November 1998.[27]

The first three launch attempts on 19, 20, and 21 November 1998 all ended in user scrubs, but the fourth countdown led to a successful launch. Apart from a 27-minute unscheduled hold for upper level winds, the countdown went smoothly, and the DELTA II lifted off the pad at 2354:00.466Z on 22 November 1998. The vehicle rolled into a flight azimuth of 97.5 degrees. The initial boost phase ended approximately 10 minutes after lift-off, and the DELTA II entered a coast phase. The next second stage burn, second/third stage separation, and third stage burn brought the vehicle downrange past the island of Guam. The spacecraft separated approximately one hour and 14 minutes after lift-off, and the mission was successful.[28]

GLOBALSTAR -1 through GLOBALSTAR -7

A total of 28 spacecraft were launched from Complex 17 on seven separate GLOBALSTAR missions between 14 February 1998 and 9 February 2000. The object of each mission was to place four commercial communications satellites in 1350-kilometer circular phasing orbits on their way to the GLOBALSTAR constellation.[29] Boeing selected a DELTA II

---

[25] Message, 45 SW Command Post to HQ AFSPC, "DELTA II/COMM Launch, 141130Z Jan 1996; 45 SW History, CY 1996, Vol I, p 91; "Delta II, shuttle, light up morning skies," *45th Space Wing Missileer,* 19 Jan 1996; Stephen Clark, "Satellite with double use launched for South Korea," *Spaceflight Now,* 22 Aug 2006.

[26] Hughes received the contract from a subsidiary of Media Most, a Russian-based private media group. The spacecraft weighed approximately 3,100 pounds at lift-off and had a mass of approximately 1,750 pounds once it began operating on-orbit. It had an on-orbit life expectancy of more than 10 years.

[27] 45 SW History, CY 1998, Vol I, p 94; Boeing, "Hughes Launches Its First Satellite for Russia," undated.

[28] 45 SW History, CY 1998, Vol I, p 94.

[29] The Italian firm Alenia Aerospazio built the spacecraft under contract to Space Systems Loral, Palo Alto, California. GLOBALSTAR was a partnership of 10 international telecommunications and aerospace companies including Loral, Qualcomm Inc., Daimler Benz Aerospace, Alcatel, China Telecom, and France Telecom.

7420 vehicle equipped with four GEMs to accomplish each mission. There was no third stage atop each launch vehicle, but a specially-designed dispenser was attached to the vehicle's second stage to inject the satellites into the proper orbital planes. Each complement of four satellites weighed approximately 3,800 pounds at lift-off, and each satellite had an on-orbit life expectancy of seven and one-half years.[30]

Virtually identical flight scenarios were followed on all seven missions. The vehicle lifted off the pad and rolled into a flight azimuth of 65 degrees. The initial boost phase ended approximately 11 minutes later, and the vehicle entered a coasting phase. Another second stage burn circularized the orbit, and it brought the payload up-range to a point where the Canberra Telemetry Tracking Station in Australia acquired and tracked both the payload and its upper stage. Two upper-tier satellites separated at 35 degrees relative to the local horizontal in-orbit plane approximately one hour and 10 minutes after lift-off. The two lower-tier satellites separated at 75 degrees relative to the local horizontal in-orbit plane about four minutes later. The dispenser remained attached to the DELTA II's second stage. Additional engine firings lowered the second stage's orbit and moved the DELTA II away from the orbital paths of the satellites.[31]

The DELTA II launch team encountered relatively few technical problems as they prepared launch vehicles for GLOBALSTAR missions, but the weather was another matter. Faced with unacceptable upper level winds on 6 February and 8 February 1998, range officials had to scrub the first two launch attempts for GLOBALSTAR-1. The weather cooperated on the third attempt a week later, and the DELTA II lifted off Pad 17A at 1434:00.170Z on 14 February 1998. Upper level wind constraints also forced officials to scrub GLOBALSTAR-2's first launch attempt on 23 April 1998, but the next countdown went smoothly – the vehicle lifted off Pad 17A at 2238:34.448Z on 24 April 1998. Two launch attempts for GLOBALSTAR-3 were scrubbed for bad weather on 8 and 9 June 1999, but conditions improved the 10th, and the launch vehicle lifted off Pad 17B without incident at 1348:43.446Z on 10 June 1999. GLOBALSTAR-4 required three countdowns before the vehicle lifted off Pad 17B at 0845:37.180Z on 10 July 1999. On the other hand, the GLOBALSTAR-5 and GLOBALSTAR-6 flights only required one countdown each. GLOBALSTAR-5 lifted off Pad 17A at 0746:03.324Z on 25 July 1999. GLOBALSTAR-6 was launched from Pad 17B at 0437:41.186Z on 17 August 1999.[32]

The launch team experienced a two-week delay in getting the second stage erected for the GLOBALSTAR-7 mission in late December 1999, but the payload was mated to the launch vehicle on 28 January 2000. Officials determined the Redundant Inertial Flight Control Assembly needed to be replaced a few days later, but that operation only took one week to complete. Only

---

GLOBALSTAR officials planned the network as a constellation of 48 operational satellites arranged in eight orbital planes at an altitude of 1400 kilometers. The planes were spaced 45 degrees apart, and six satellites were assigned to each orbital plane to support global voice and data communications throughout the network.

[30] 45 SW History, CY 1998, Vol I, pp 90, 91; 45 SW History, CY 1999, Vol I, pp 77, 80, 81, 82; 45 SW History, CY 2000, Vol I, p 107; Fact Sheet, "GLOBALSTAR," *Small Satellites Home Page*, undated; Fact Sheet, Globalstar, "How Globalstar Works," 2009.

[31] 45 SW History, CY 1998, Vol I, pp 90, 91, 92; 45 SW History, CY 1999, Vol I, pp 77, 80, 81, 82; 45 SW History, CY 2000, Vol I p 107.

[32] 45 SW History, CY 1998, Vol I, pp 91, 92; 45 SW History, CY 1999, Vol I, pp 79, 80, 81, 82, 83.

one countdown was required to launch GLOBALSTAR-7, and the DELTA II lifted off Pad 17B at 2124:00.297Z on 8 February 2000. The mission was successful.[33]

Other GLOBALSTAR payloads were launched successfully on Soyuz-Ikar boosters from Baikonur Cosmodrome in Kazakhstan in 1998 and 1999, and the entire first-generation GLOBALSTAR constellation was on-orbit by early 2000. The constellation proved to be so successful that Globalstar, Inc. signed a contract with Alcatel Alenia Space in December 2006 to design, build and deliver the next generation of 48 GLOBALSTAR satellites. Four GLOBALSTAR spacecraft were launched aboard a Russian Soyuz-FG vehicle from Baikonur in late May 2007.[34]

Some of the first-general GLOBALSTAR satellites were nearing the end of their useful lives in late 2005. In October 2005, GLOBALSTAR officials advised the Federal Communications Commission that GLOBALSTAR planned to transfer some of its old satellites to higher "disposal" orbits (e.g., circular orbits between 1,515 kilometers and 2,000 kilometers above Earth) in the near future. Four of the satellites were relocated to disposal orbits before the end of 2005.[35]

DELTA III Missions

Three DELTA III missions were launched from Pad 17B between 27 August 1998 and 24 August 2000. All three were prompted by a growing demand for lower-cost, highly reliable launch services. McDonnell Douglas began developing the DELTA III in the spring of 1995, and Boeing completed the program after it bought McDonnell Douglas on 4 August 1997. The goal was a more capable, more economical, and simpler launch vehicle with reliability as least as good as the DELTA II. To meet its customers' requirements, the DELTA III needed more than double the payload lift capability of the DELTA II (e.g., 8,400 pounds to geosynchronous transfer orbit versus 4,120 pounds for the DELTA II).[36] The contractor believed time and money could be saved if: 1) parts counts were minimized, 2) DELTA II/III parts commonality was maximized, and 3) passive systems were substituted for active systems wherever possible.[37]

With those objectives in mind, DELTA III engineers chose the upgraded flight safety and avionics equipment developed for and used on the DELTA II. The DELTA III's attitude control

---

[33] 45 SW History, CY 2000, Vol I, pp 107, 108.

[34] Fact Sheet, "GLOBALSTAR," *Small Satellites Home Page*, undated; News Release, Globalstar, Inc., "Globalstar signs contract for new satellite constellation," 5 Dec 2006; NASA, "Soyuz-FG launches with four Globalstar satellites," 29 May 2007.

[35] NASA, "Disposal of Globalstar Satellites" *Orbital Debris Quarterly News,* Vol 10, Issue 1, Jan 2006.

[36] In May 1995, McDonnell Douglas had a firm order for 10 DELTA III launches and options for 10 more from Hughes Space and Communications Company. Hughes needed the launch services for its HS-601 series satellites, which weighed more than 8,000 pounds apiece at lift-off. By early 1998, the order had grown to 13 launches for Hughes and five more DELTA III launches for Space Systems Loral. The GALAXY X spacecraft carried on the first DELTA III weighed approximately 8,546 pounds at lift-off, and the spacecraft would have retained a mass of 4,663 pounds in geosynchronous orbit (had the flight been successful). The GALAXY X was much too heavy for the DELTA II to lift into geosynchronous transfer orbit.

[37] Handout, the Boeing Company, "DELTA III – The Commercial Evolution of DELTA," 3 Apr 1998; Backgrounder, the Boeing Company, "Boeing Expendable Launch Systems," undated; "Boeing Delivers First Delta 3 Booster," *Aviation Week & Space Technology,* 23 Feb 1998; 45 SW History, CY 1997, Vol I, p 50.

system was identical to the DELTA II's attitude control system, and the majority of ordnance carried on the DELTA II was shared by the DELTA III. Both vehicles used identical liquid oxygen/liquid hydrogen propulsion systems on the first stage, but the DELTA III's first stage fuel tank was increased to 13.1 feet in diameter to complement the second stage's greater girth and increase flight stability and tolerance of upper level winds. The DELTA III was a thicker, heavier vehicle than the DELTA II, but it was not be an appreciably taller one. The contractor's design strategy ensured DELTA III and DELTA II vehicles could be processed on Pad 17B interchangeably. It also reduced the number and complexity of facility modifications to Pad 17B's Mobile Service Tower and launch mount in 1997 and 1998.[38]

There were, or course, significant differences between the DELTA III and its predecessor. The DELTA III was equipped with stretched, enlarged versions of the Alliant Techsystems Graphite Epoxy Motors (GEMs) found on the DELTA II. The new Large Diameter Extended Length Graphite Epoxy Motors (GEM-LDXLs) were nearly six inches wider and four feet longer than the DELTA II's GEMs, and three of the DELTA III's GEMs incorporated a Thrust Vector Control (TVC) system with flex nozzles to facilitate the flight after lift-off. The new GEMs were about 25 percent more powerful than the old GEMs. They each produced an average thrust of 137,362 pounds at lift-off and 142,242 pounds at altitude. The DELTA III carried a cryogenic (liquid oxygen/hydrogen) Pratt & Whitney RL 10B-2 engine in its second stage instead of the DELTA II's hypergolic Aerojet AJ10-118K engine. Not only was the RL 10B-2 more than twice as powerful as the Aerojet engine, it also eliminated any need for a third stage and the storage of hazardous propellants (Aerozine 50 and nitrogen tetroxide). The DELTA III's fairing was enlarged to 13.1 feet in diameter (versus the DELTA II's 9.5-foot or 10-foot-diameter fairing), and it was made of composites to lower the cost, the number of parts, and the weight of the vehicle. With the exception of cryogenic tankage and its attachment rings, all primary structures on the DELTA III were made of composite materials. Another simplification in the DELTA III design was the use of robust, separate, self-supporting liquid oxygen and liquid hydrogen tanks in lieu of an arrangement with a common bulkhead. Boeing built both oxidizer tanks for the DELTA III. Mitsubishi Heavy Industries built the hydrogen tanks for the first and second stages.[39]

The DELTA III was built mainly in Boeing's plant in Huntington Beach, California, before it was assembled in Boeing's facility in Pueblo, Colorado. Boeing accomplished vehicle and payload integration at the launch site. Pad 17B was upgraded in 1997 and 1998 to handle DELTA III and DELTA II vehicles equally well, thereby reducing infrastructure costs and making the best use of vehicle integration and launch team personnel.[40]

Eastern Range support for DELTA III missions was similar to the instrumentation requirements for DELTA II flights. Range instrumentation for DELTA III missions included radar assets on the Cape, Merritt Island, Jonathan Dickinson, and the range stations on Antigua and Ascension. The Cape, Jonathan Dickinson and Antigua provided command/destruct functions.

---

[38] Handout, the Boeing Company, "DELTA III – The Commercial Evolution of DELTA," 3 Apr 1998; Backgrounder, the Boeing Company, "Boeing Expendable Launch Systems," undated; "Boeing Delivers First Delta 3 Booster," *Aviation Week & Space Technology,* 23 Feb 1998; Summary, ANSER, "A Historical Look at United States Launch Vehicles, 1967-Present," Jan 1997, pp III.A-11, III.A-12; 45 SW History, CY 1997, Vol I, p 50.

[39] Handout, the Boeing Company, "DELTA III – The Commercial Evolution of DELTA," 3 Apr 1998; "Boeing Delivers First Delta 3 Booster," *Aviation Week & Space Technology,* 23 Feb 1998, p 95; 45 SW History, CY 2000, Vol I, p 117.

[40] Handout, the Boeing Company, "DELTA III – The Commercial Evolution of DELTA," 3 Apr 1998.

Telemetry support came from systems on Merritt Island, Antigua and Ascension. Optical support included the Distant Object Attitude Measuring System (DOAMS) at Cocoa Beach, optical systems at the Cape and KSC, and the Fixed IGOR at Patrick AFB. A full array of computers, weather services, HH-60G helicopters, a weather reconnaissance Learjet, and off-range resources rounded out the instrumentation assets covering DELTA III spaceflights.[41]

The object of the first DELTA III mission was to place the GALAXY X commercial communications spacecraft into a nominal transfer orbit. Hughes Space and Communications Company built the GALAXY X spacecraft for the PanAmSat Corporation. The spacecraft was an HS 601HP model based on the world's best-selling large satellite series which Hughes introduced as its HS 601 line in 1987. The GALAXY X was equipped with gallium arsenide solar panels to provide 5.8 kilowatts of power in geostationary orbit. The spacecraft carried 24 C-Band transponders and 24 Ku-Band transponders to provide digital and analog telecommunications throughout the continental United States and the Caribbean. The spacecraft (in stowed configuration) measured 19 feet two inches long by 8 feet 10 inches wide by 11 feet 9 inches deep. As advertised, it weighed 8,546 pounds at lift-off and 4,663 pounds on orbit. It had an on-orbit life expectancy of 12 years.[42]

Technicians and engineers erected the DELTA III's first stage on Pad 17B on 4 May 1998. They added the last of the vehicle's GEMs on 25 July, and they erected the second stage on 27 July 1998. Following a flight simulation on 7 August and a Wet Dress Rehearsal (WDR) on 12 August, the contractor mated the GALAXY X spacecraft to the vehicle on 13 August 1998. The countdown on 26 August 1998 included two unscheduled holds, but the DELTA III lifted off the launch pad at 0117:00.205Z on 27 August 1998.[43]

Between 55 and 65 seconds into the flight, roll oscillations around 4 Hertz (4 Hz) prompted the launch vehicle's control system to gimbal its three swiveling GEMs to compensate for the oscillations until the hydraulic system ran out of fluid.[44] At 65 seconds, the GEMs ceased to swivel, and two of them were stuck in positions that helped overturn the vehicle. The DELTA III's main engine gimbaled to correct the overturning movement, but, as it fought against its big, 13.1-foot-diameter fairing and its GEMs, it quickly lost the battle. The vehicle yawed about 35 degrees, and it began to disintegrate at an estimated altitude of 60,000 feet about 71 seconds after lift-off. In accordance with safety guidelines, the Mission Flight Control Officer on duty sent destruct functions 75 seconds into the flight, and that action completed the destruction of the vehicle. All debris fell into the Atlantic Ocean between 10 and 15 miles east of Cape Canaveral. According to Air Force safety officials, the mishap had minimal affect on aquatic life in the area. Most of the DELTA III's propellants were consumed in the explosion following destruct actions,

---

[41] 45 SW History, CY 1998, Vol I, p 99; 45 SW History, CY 1999, Vol I, p 87; 45 SW History, CY 2000, Vol I, p 116.

[42] 45 SW History, CY 1998, Vol I, p 99.

[43] 45 SW History, CY 1998, Vol I, p 100.

[44] The development of a closed loop hydraulic system for the swiveling GEMs was discarded in the design phase because it would have added too much weight and cost to the vehicle. Had the vector control system not responded to the 4 Hz oscillations, there would have been enough hydraulic fluid in the reservoir to fly two and one-half normal missions.

and the remaining aerosols dissipated to "essentially undetectable" levels before they reached the ocean surface.[45]

The DELTA III and its payload were insured for $225 million, so the failure was no small affair. Moreover, Boeing had to restore its customers' confidence in the DELTA III before the next flight, so the company chose reviewing officials who were not part of the original DELTA III review team to ensure an objective inquiry. Hughes, the Federal Aviation Administration, NASA, the Air Force, the Aerospace Corporation, and Alliant Techsystems also supported Boeing's investigation team.[46]

In the course of its mishap investigation, Boeing conducted a two-week revalidation process for every DELTA III system. Analysis revealed that the DELTA III's aerodynamics and GEM functions were rather different from the DELTA II. Put simply, the DELTA III's main engine did not have enough power to maneuver the vehicle during the first 81 seconds of flight if the swiveling GEMs hindered the operation. During the flight on 27 August, a 3-Hertz roll instability in the DELTA III's air-lit GEMs developed into the dominant 4-Hertz oscillation as the aft end of the DELTA III burned propellant and became much lighter. The vehicle's control system software commanded the Thrust Vector Control system to respond to the oscillation. Ironically, this only made things worse. Once all the hydraulic fluid was expended, the *oscillation smoothed out*. As noted, two of the three swiveling GEMs were stuck in the wrong position, and wind shear forced the DELTA III to yaw and break up seven seconds later. The review team concluded that the flight would not have failed if the control system software hadn't commanded the TVC system to respond to the 4 Hertz oscillation. It would have smoothed out on its own. As a result of the investigation, Boeing decided to change an instruction to the flight control system so it would *identify and ignore* the 4-Hertz roll oscillation in subsequent DELTA III flights. Boeing planned to have the next DELTA III ready for another commercial flight in early 1999.[47]

The second DELTA III mission was launched in early May 1999, and it featured another one of Hughes' HS 610 spacecraft – the ORION-3. The Loral Corporation bought the spacecraft to provide television, high-speed voice, and data communications to Korea, China, India, Japan, Australia, Southeast Asia, Oceania and Hawaii. The satellite had an operational life expectancy of 12 years, and its intended destination was a geosynchronous orbit above 139 degrees East longitude.[48]

Technicians and engineers erected the DELTA III's first stage on Pad 17B on 20 January 1999. The vehicle's GEMs should have been hoisted into place a few days later, but the crane used to hoist the GEMs required repairs. It was not back in service until 9 February 1999. In the meantime, Boeing began having doubts about the reliability of the second stage engine (engine #802), so the company replaced it with engine #803 in early February 1999. Technicians added the last of the GEMs to the vehicle on 10 February, and engineers erected the second stage on 6 March 1999. In a letter dated 11 March 1999, Air Force Space Command's Director of

[45] 45 SW History, CY 1998, Vol I, pp 100, 102.

[46] 45 SW History, CY 1998, Vol I, pp 102.

[47] News Release, the Boeing Company, "Boeing Pinpoints Cause of Delta III Failure, Predicts Timely Return to Flight," 5 Sep 1998; "Boeing Delta 3 Explodes; Commercial Debut Ruined," *Aviation Week & Space Technology,* 31 Aug 1998; "Delta 3 Roll Mode To Be Locked Out of Control System," *Aviation Week & Space Technology,* 2 Nov 1998.

[48] 45 SW History, CY 1999, Vol I, p 87.

Operations, Major General Robert C. Hinson, agreed with the 45th Space Wing Commander's decision to approve the DELTA III for its "return to flight." Following a flight simulation on 19 March and a Wet Dress Rehearsal (WDR) on 22 March, the contractor mated the ORION-3 spacecraft to the vehicle on 27 March 1999.[49]

Five launch attempts were required to send the ORION-3 on its way. During the first countdown on 6 April 1999, the temperature, wind speed and direction of the wind exceeded toxic limits throughout the launch window, so officials scrubbed the launch on that date. The second countdown on 7 April included three unscheduled holds, and officials finally scrubbed the launch for an instrumentation problem on that date. Officials scrubbed the third launch attempt on 22 April for a popped circuit breaker that would not reset. The countdown on 23 April was scrubbed for a „hangfire,' but the fifth launch attempt was successful. There were no unscheduled holds during the countdown, and the DELTA III lifted off the launch pad at 0100:00.165Z on 5 May 1999. Unfortunately, the mission failed.[50]

Early reports on the mishap indicated the vehicle's first stage and GEMs performed properly. The second stage's first burn placed the payload in a proper preliminary orbit (e.g., 85 x 744 nautical miles), but when the second stage fired again – in this instance, about 22 minutes after lift-off — telemetry reported a "shock" about 3.5 seconds into the burn. Then the second stage simply quit firing. Though the payload separated from the second stage as planned, it was left in a useless low-Earth orbit.[51]

The spacecraft and its DELTA III booster were fully insured for $265 million, but Boeing was worried about future launches and future customers. There had been no successes since the DELTA III's introduction in August 1998, and the mishap was the second failure in a row for the new launch vehicle. An investigation board was convened almost immediately. Dr. Russell Reck, Boeing's Director of Engineering Technology for Expendable Launch Systems, led the investigation board. The effort was supported by Hughes, Loral, Pratt & Whitney, Lockheed Martin, NASA, the Air Force and the Federal Aviation Administration. Since the mishap centered on the second stage's engine, Pratt & Whitney grounded all versions of its RL10 engine in May 1999 including the ones it supplied for Lockheed Martin's ATLAS IIAS/CENTAUR vehicles. The ban on RL10s was lifted in August 1999, but it would take Pratt & Whitney a long time to live down the DELTA III ORION-3 mishap.[52]

Boeing launched its final DELTA III in late August 2000. The object of the mission was to place the DM-F3 simulated payload into a nominal transfer orbit to verify the reliability of the DELTA III launch system. The spool-shaped DM-F3 was designed to have the same center of gravity and structural frequency response as the 9,480-pound ORION-3 spacecraft lost on the

[49] 45 SW History, CY 1999, Vol I, p 88.

[50] 45 SW History, CY 1999, Vol I, p 88.

[51] "DELTA III FAILS TO PLACE ORION 3 IN PROPER ORBIT," *Florida Today* Space Online, 5 May 1999; "Delta III Failure Ruins Comeback, Strands Orion," *Aviation Week & Space Technology,* 10 May 1999.

[52] "Delta III Failure Ruins Comeback, Strands Orion," *Aviation Week & Space Technology,* 10 May 1999; "Grounding of RL10 Engine Threatens More Launches," *Space News,* 24 May 1999; "RL10 To Undergo Ultrasound Inspections," *Space News,* 9 Aug 1999.

second DELTA III flight failure in May 1999.[53] The DM-F3 mission was essentially a repeat of the ORION-3 mission with a dummy payload.[54]

Engineers and technicians erected the DELTA III's first stage on Pad 17B on 26 June 2000. Nine GEM-LDXLs should have been hoisted into place a few days later, but foul weather and lightning delayed that operation until 6 July 2000. Officials delayed the second stage's erection from 19 July to 24 July to complete a Vehicle Readiness Review (VRR). They also delayed the Simulation Flight until 7 August 2000, but the change in schedule had no impact on the remaining milestones. The Wet Dress Rehearsal was completed successfully on 9 August, and the contractor mated the DM-F3 to the launch vehicle on 14 August 2000. Preparations continued for the launch, which was scheduled for 23 August 2000.[55]

The range remained 'Green' throughout the countdown on the 23rd. Boeing detected a temperature out-of-tolerance reading on the DELTA III second stage shortly before the launch, but the indication cleared after five minutes. The user was 'go for launch' at 1057Z. The DELTA III lifted off the launch pad at 1105:02.050Z on 23 August 2000.[56]

The flight went as planned. The first six GEM-LDXLs burned out at T plus 77.5 seconds, and the remaining three GEM-LDXLs ignited about 1.5 seconds later. After GEM-LDXL burnout around T plus 157 seconds, the final three empty GEM-LDXL cases were jettisoned around T plus 160 seconds. Following Main Engine and Vernier Engine cut-off, the first stage separated from the rest of the vehicle around T plus four minutes and 30 seconds. The second stage ignited around T plus four minutes and 41 seconds, and it continued to thrust until about T plus 13 minutes and 37 seconds. The second stage restarted around T plus 21 minutes and 35 seconds. Ascension tracked the second stage restart as well as the upper stage's shutdown. The DM-F3 payload separated from the vehicle about 36 minutes after lift-off, near the apogee of the transfer orbit.[57]

Boeing was very pleased with the results of the mission. Flight data confirmed the successful operation of all DELTA III systems, and the DM-F3 payload separated from the vehicle as expected. Because the mission was flown until all propellant was exhausted, the flight apogee varied from 10,935 nautical miles to 15,410 nautical miles. The payload was placed within that range at an apogee of 11,156 nautical miles. Had there been an actual functioning satellite onboard the vehicle, the mission would have met all customer requirements.[58]

---

[53] Boeing attempted to attract a "revenue generating payload" for the flight for several months, but the market was slow, and there were no prospects. The company eventually decided to absorb $85 million in costs to demonstrate the DELTA III's reliability and maintain its "market viability." Unfortunately, there were no customers for the DELTA III later on, and Boeing had to refocus its commercial energies on its old DELTA II and its new DELTA IV line.

[54] 45 SW History, CY 2000, Vol I, pp 116, 117; "Boeing Sets Costly Delta III Flight Test," *Aviation Week & Space Technology,* 19 Jun 2000.

[55] 45 SW History, CY 2000, Vol I, p 117

[56] 45 SW History, CY 2000, Vol I, p 118.

[57] 45 SW History, CY 2000, Vol I, p 117; "Boeing's Delta 3 rocket finally flies with success," *Florida Today,* 24 Aug 2000.

[58] 45 SW History, CY 2000, Vol I, p 118; "We Have Liftoff! Delta III rocket blasts off to success," *Inside Delta* (Boeing Company Publication), Vol 4, No 9, Sep 2000.

CHAPTER V

1 SLS INACTIVATION & TRANSFER OF DELTA II FACILITIES

On 28 March 2003, Headquarters Air Force Space Command issued Programming Message 03-02, "Deactivation of ATLAS II/III, DELTA II and TITAN IV Space Launch Systems." Among other things, the message provided guidance for the deactivation of DELTA II launch facilities and infrastructure following the last Air Force-sponsored DELTA II mission at Cape Canaveral. In response to Programming Message 03-02, 45th Space Wing officials formed the DELTA II Deactivation Working Group under the co-chairmanship of the 45th Space Wing Plans Office (45 SW/XP) and the 1st Space Launch Squadron (1 SLS). The Group had the job of planning DELTA II deactivation activities and separating the Air Force from any further financial liability for DELTA II operations at Cape Canaveral once the DELTA II/GPS IIR-21 mission was completed.[1]

As the GPS IIR-21 launch drew closer, officials drafted a 45th Space Wing implementation plan entitled, "Deactivation of Delta II Space Launch Systems," in the fall of 2007. The plan offered details concerning the deactivation and the transfer of DELTA II launch infrastructure at Cape Canaveral and the inactivation of launch operations and supporting units, including the 1 SLS. The 45th Space Wing Commander delegated responsibility and authority for the DELTA II deactivation to the 45th Space Wing Plans Office. Mr. Byron G. Whiteman was designated the primary I-Plan manager, and annex managers were assigned from various organizations to troubleshoot problems and ensure completion of required tasks.[2]

On 15 November 2007, Mr. Whiteman and Major Tim Spies conducted a briefing to provide an overview for the DELTA II deactivation for Brigadier General Susan J. Helms and other 45th Space Wing leaders at Patrick AFB. According to Mr. Whiteman, three deactivation stages had been identified in a Headquarters Air Force Space Command programming message in 2003. The stages began with normal operations (Phase I), continued through a transition (Phase II), and concluded with an end state (Phase III). The boundaries between the phases were flexible, and delays in one phase could cause 'waterfall' effects on the next phase or phases. On the other hand, some actions within different phases could be accomplished in parallel to achieve results in a more expeditious manner.[3]

Headquarters Air Force Space Command changed its guidance on the DELTA II deactivation on 15 April 2008. General C. Robert Kehler announced the Air Force would be transferring DELTA II facilities and infrastructure to NASA to support that civilian agency's Gravity Recovery and Interior Laboratory (GRAIL) in 2011. Air Force Space Command followed up its guidance with a programming message, "Deactivation of the Delta II Space Launch System at the 45th Space Wing," dated 8 December 2008. The new guidance required a re-write of the

---

[1] Discussion, M. Cleary with Mr. Byron G. Whiteman, 45 SW/XPR, 15 Dec 2008; E-Mail, Mr. Byron G. Whiteman, 45 SW/XPR to Mark C. Cleary, 45 SW/HO, "RE: Historical Info on Complex 17 Transfer – Implementation Plan 08-02," 15 Dec 2008.

[2] Excerpt, 45 SW/XPR, "Delta II Deactivation IPlan Tasks (Draft)," Sharepoint.patrick.af.mil, Delta II Deactivation Effort Website, 18 Sep 2008.

[3] Slide Briefing, Mr. Byron G. Whiteman, 45 SW/XPR, and Major Tim Spies, 1 SLS, "Delta II Deactivation," briefed 15 Nov 2007.

implementation plan mentioned earlier. Now the DELTA II Deactivation Plan focused on preserving DELTA II launch infrastructure to ensure a smooth transfer of launch assets to NASA. Nevertheless, considerable work would be required to: 1) inactivate the 1 SLS, 2) develop a memorandum of agreement with NASA's Kennedy Space Center to implement the transfer, and 3) consolidate and close out equipment accounts. Uncertainty surrounded the launch date for the GPS IIR-21 mission, but officials completed revisions to the implementation plan by the end of 2008.[4]

Most tasks involving manpower issues were completed by the fall of 2008. The 45th Launch Group was busy identifying equipment and consolidating equipment accounts for closeout, and the Memorandum of Agreement (MOA) between the Air Force and NASA had been drafted. In the meantime, Mr. Whiteman established a SharePoint site to track progress/completion of specific DELTA II deactivation tasks and to address other issues associated with the DELTA II program transition.[5]

It appeared by early June 2009 that the GPS IIR-21 mission might be launched from Complex 17 sometime in August 2009. One additional complication in the transfer was the U.S. Missile Defense Agency's Space Tracking and Surveillance System (STSS) demonstration satellite mission, which was tentatively scheduled to lift off a few weeks after GPS IIR-21.[6] Though the STSS demonstration flight was clearly a military mission, NASA arranged the DELTA II launch service for MDA.[7]

In any event, the Air Force expected to transfer many of its DELTA II facilities at Cape Canaveral to NASA after the GPS IIR-21 and the STSS demonstration missions were launched. Those facilities would probably include: 1) Complex 17's blockhouse, 2) both launch pads, 3) two Mobile Service Towers (MSTs), 4) nearly a dozen camera towers, 5) storage facilities for fuel, liquid oxygen, and liquid nitrogen, 6) some administration buildings,[8] and 7) various electrical, water, and sewage/sanitation facilities. Thirty days would be needed to complete normal pad clean-up operations after the last launch, but facilities could be transferred to NASA as each one became ready for transfer. The entire operation would probably take several months. The latest

---

[4] Excerpt, 45 SW/XPR, "Delta II Deactivation IPlan Tasks (Draft)," *Sharepoint.patrick.af.mil*, Delta II Deactivation Effort Website, 18 Sep 2008; Message, HQ AFSPC to 14 AF et al, "Headquarters Air Force Space Command (HQ AFSPC) Programming Message (PMSG) 08-09: Deactivation of the DELTA II Space Launch System at the 45th Space Wing," 8 Dec 2008; Discussion, M. Cleary with Mr. Byron G. Whiteman, 15 Dec 2008.

[5] Excerpt, 45 SW/XPR, "Delta II Deactivation IPlan Tasks (Draft)," Sharepoint.patrick.af.mil, Delta II Deactivation Effort Website, 18 Sep 2008; Discussion, M. Cleary with Mr. Byron G. Whiteman, 15 Dec 2008.

[6] In early May 2009, the U.S. Missile Defense Agency (MDA) announced that one of two STSS satellites had been shipped to Cape Canaveral in anticipation of the STSS demonstration flight from Complex 17. Once the second STSS satellite arrived to complete the tandem payload, the STSS demonstration flight might be launched within weeks of the GPS IIR-21 mission.

[7] Discussion, M. Cleary with Mr. Byron Whiteman, 8 Jun 2008; News Release, MDA, "Space Tracking and Surveillance System's Demonstration Program Space Vehicle Two Ships to Canaveral, 4 May 2009; Slide Briefing, Mr. Byron Whiteman, 45 SW/XPR, "Delta II Deactivation," 3 Jun 2009.

[8] The DELTA Operations Building (formerly known as the 1 SLS Operations Building) was not included among the buildings being considered for transfer to NASA, though three rooms within the building might be offered for NASA's use.

version of the draft MOA was awaiting signature in early June 2009,[9] but it could not be released publicly at that time.[10]

The transfer of DELTA II facilities to NASA and the inactivation of the 1 SLS will be remembered as major milestones in Cape Canaveral's history. In many ways, they are the end of a long and interesting story. More than half a century earlier, the Cold War was in full swing when the Air Research and Development Center (ARDC) ordered development of the THOR (Weapon System 315A) program "as soon as possible." The Wright Air Development Center sponsored the missile initially, and the Air Force Missile Test Center (AFMTC) hosted the THOR at Cape Canaveral. Following a series of meetings between AFMTC and Western Development Division officials in February and March 1955, support requirements were worked out for two launch pads, a blockhouse, a missile guidance site, one service stand, airborne guidance test equipment, housing and messing facilities.[11]

The THOR was given equal priority with the ATLAS in December 1955, and the Western Development Division became the missile's new sponsor at that time. The first THOR test missile was launched in January 1957, and, by that time, THOR launching facilities consisted of a blockhouse, one launching pad with a service tower (Pad 17B), a second partly-finished launch pad (17A), and a 40,000-square-foot assembly building (Hangar M). The principal contractors were Douglas Aircraft Company for the airframe, Bell Telephone Laboratories for the radio-inertial guidance system, A.C. Spark Plug for the more advanced all-inertial guidance system, General Electric for the nose cone, and North American Aviation for the rocket motors. The THOR weighed 110,400 pounds, and it was 62.5 feet long and 8 feet in diameter. It was propelled by a single rocket motor rated between 135,000 pounds and 150,000 pounds of thrust.[12]

The first THOR was launched from Cape Canaveral on 25 January 1957. Engine ignition and main stage operation were normal at launch, but a liquid oxygen valve failed almost immediately after lift-off, and the missile slipped back through Pad 17B's launcher ring to explode on the deflector plate below. A low order explosion and fire destroyed the missile and damaged the pad, and the launch failure delayed the next launch until mid-April 1957. The second THOR launch on 19 April was more successful, but a third missile (Number 103) exploded on the pad on 21 May after its fuel tank ruptured five minutes before its intended launch. Pad 17B had to be refurbished once again, but Pad 17A was completed in July 1957, so it was used for the fourth THOR on 30 August 1957. That lift-off was successful, but the missile broke in half 93 seconds into the flight and plunged into the Atlantic about 20 miles from the launch site. The fifth THOR was launched from Pad 17B on 20 September 1957, and it met all its test objectives.[13]

---

[9] There were actually two MOAs – the one awaiting signature would be signed at the Air Force/NASA Headquarters level. The other MOA was 'in draft' in early June 2009, and it would be signed at the 45th Space Wing/Kennedy Space Center level.

[10] Discussion, M. Cleary with Mr. Byron Whiteman, 15 Jun 2008; Summary, Mr. Byron Whiteman, 45 SW/XPR, "NASA Permitted – Post Grail – Delta II Deactivation Facility Demolition List," 8 Jun 2009.

[11] 45 SW History Office, "The 6555th: Missile and Space Launches through 1970," p 96.

[12] *Ibid.*

[13] 45 SW History Office, "The 6555th: Missile and Space Launches through 1970," p 101.

Unfortunately, Pad 17A received its own baptism in fire on 3 October 1957 when the sixth THOR lost thrust, fell back through the launching ring, and burned. However, two more THOR launches on 11 October and 24 October were successful, and they concluded the new missile's airframe/propulsion testing phase at the Cape. Guidance system tests began with a partially successful THOR flight from Pad 17B on 7 December 1957. A subsequent flight from Pad 17A tested the THOR's all-inertial guidance system on 19 December 1957. It was completely successful.[14]

THOR nose cone separation tests began with a flight from Pad 17B on 28 February 1958, but the second launch in that series ended in another explosion and fire on Pad 17B on 19 April. The third nose cone test was flown successfully on 13 June, and a THOR tactical launcher performed well during a guidance system test launch offsite (e.g., on Pad 18B) on 4 June 1958. Test flights to validate overall refinements in the THOR got underway on 5 November 1958, and five of those flights were launched from Pad 17B or Pad 18B by the end of the year. Three of the five met their test objectives, and so did eight of the next nine THORs launched from 17B or 18B between 30 January and the end of June 1959. An 82-foot-long variant of the THOR known as the THOR-ABLE was used in a dozen flights from Pad 17A between 23 April 1958 and 12 June 1959 to test two different ablative nose cone designs for the ATLAS program. Eleven out of 14 THORs met their flight test objectives in the last half of 1959.[15]

Contractors launched many THORs at Cape Canaveral, but the Air Force conducted its ballistic missile Combat Training Launch (CTL) operations at Vandenberg AFB, California. The first missile launched from Vandenberg was also the first THOR CTL operation, and a Strategic Air Command (SAC) missile crew completed it successfully on 16 December 1958. The British participated in THOR launches at Vandenberg AFB and Cape Canaveral a little later on, and they increased their presence at Vandenberg and the Cape as blue suit and contractor support steadily diminished over the next two and one-half years.[16]

As testing continued at the Cape, THOR missiles began to arrive in the United Kingdom in September 1958. Sixty THOR launch sites (assigned to four THOR squadrons) went on alert between June 1959 and April 1960. THOR launch operations were performed exclusively by Royal Air Force personnel after June 1961. Since the THOR's mission could be assumed by other weapon systems after 1962, the THORs were pulled out of Great Britain between November 1962 and August 1963. The missiles were returned to the United States. The THOR was matched with several different high-energy upper stages in the late 1950s,[17] and later versions of the old booster were employed as the first stage of DELTA and DELTA II launch vehicles used for military, civilian agency, and commercial satellite missions.[18]

---

[14] 45 SW History Office, "The 6555th: Missile and Space Launches through 1970," p 101.

[15] 45 SW History Office, "The 6555th: Missile and Space Launches through 1970," p 106.

[16] 45 SW History Office, "The 6555th: Missile and Space Launches through 1970," pp 107, 108.

[17] One outstanding example was the THOR-ABLE, which was used on a dozen flights from Pad 17A between 23 April 1958 and 12 June 1959 to test two different nose cone designs for the ATLAS missile program. The ABLE upper stage was a modified Aerojet-General booster rated at 7,700 pound of thrust. The nose cones, built by General Electric and AVCO, were designed to absorb the intense heat of atmospheric reentry by shedding thin layers of their surfaces.

[18] 45 SW History Office, "The 6555th: Missile and Space Launches through 1970," pp 106, 108.

Many important scientific payloads, communications spacecraft, and weather satellites were launched from Complex 17 before the DELTA II program got underway in the late 1980s. Some of the most important ones were:

- ECHO – A series of inflatable passive communications satellites.

- RELAY – A pair of experimental communications satellites sponsored by NASA to demonstrate the spacecraft's ability to actively relay telephone, telegraph, facsimile and high-quality wideband television communications between continents.

- EXPLORER – A series of research satellites launched to investigate energy fields and energetic particles having an effect on communications satellites.

- OSO – A series of Orbiting Solar Observatories designed to monitor the Sun's activity and its production of harmful protons.

- PIONEER – A series of interplanetary spacecraft designed to gather data on plasma, energetic particles, and magnetic fields propagated by the Sun.

- TIROS – A series of Television Infrared Observation Satellites. The first eight TIROS spacecraft provided valuable weather data and pushed the state of the art in satellite weather forecasting and weather satellite design in the early 1960s.

- GOES – A series of Geostationary Operational Environmental Satellites designed to provide worldwide weather coverage from space.

- INTELSAT I, II and III – The first three series of communications satellites sponsored by the International Telecommunications Satellite Consortium. The satellites revolutionized television coverage by allowing worldwide broadcasts of events as they unfolded "live."

In all, 322 major missions were launched from Complex 17 between January 1957 and the end of March 2009. That's nearly 10 percent of all the missile and space flights supported by the Eastern Range since July 1950 (3,390 as of this writing). In the past 21 years, 358 major space missions have been launched from Cape Canaveral and the Kennedy Space Center. Of those, 109 were launched on DELTA II and DELTA III space launch vehicles.[19]

The Air Force has not been shy about giving credit where credit is due. Following the introduction of the DELTA II in 1989, the Eastern Space and Missile Center received two Air Force Excellence Awards for three consecutive years of service ending on 30 September 1991. Following ESMC's transformation into the 45th Space Wing in November 1991, the Wing and its subordinate units received no fewer than a dozen Air Force Outstanding Unit Awards for meritorious achievement or meritorious service. The latest award was conferred for service from 1 December 2006 through 30 September 2008.[20]

---

[19] 45 SW History Office, "Eastern Range Launch Database," updated 24 May 2009; 45 SW History Office, "Space Launch Numbers, 1958 -2009."

[20] Summary, 45 SW History Office, "Honors," updated Mar 2009.

The 1 SLS should be inactivated by the fall of 2009, but that event might not be the final chapter in the 1 SLS' career. The Air Force can either inactivate or deactivate a unit, but there is a crucial difference between the two actions: deactivated units are retired from service permanently, but inactivated units can be *re-activated*. We at the 45th Space Wing witnessed this process first-hand when the 5th Space Launch Squadron (5 SLS) came and went, only to come back again.

Following its first four years of service, the 5 SLS was inactivated, and its resources were absorbed by the 3rd Space Launch Squadron on 29 June 1998. The reorganization was prompted by an Air Force Space Command initiative to consolidate space launch squadrons at Cape Canaveral and Vandenberg AFB in 1998 and 1999.[21] Five years later, the 5 SLS was re-activated and assigned to the newly activated 45th Launch Group to provide oversight for DELTA IV and ATLAS V operations at Cape Canaveral. (The 5 SLS reactivation ceremony was held in Boeing's Horizontal Integration Facility at Cape Canaveral on 5 December 2003.) Our experience with the 5 SLS demonstrated that a unit inactivation may not be permanent. It remains to be seen what will happen to the 1 SLS in later years.[22]

No matter what the future holds, the men and women of the 1 SLS can take special pride in having participated in a great adventure. The future remains hidden, but I suspect that decades from now someone somewhere will be heard saying, "I'll bet those old launch teams had some really fascinating experiences; you know – stories that never made it into the history books. Back around the turn of the 21st Century, launch operations were *really* exciting – anything could happen. It must have been something to work on so many different kinds of payloads and launch all those old rockets!"

---

[21] The initiative was approved in a memorandum dated 27 May 1998. Inactivation of the 4th Space Launch Squadron (4 SLS) at Vandenberg AFB was also approved under the memorandum, and that squadron's resources were transferred to the 2nd Space Launch Squadron (2 SLS).

[22] 45 SW History, CY 1998, Vol I, p 1; 45 SW History, CY 2003, Vol I, pp 2, 3.

APPENDIX A

LISTS OF COMMANDERS

(45TH SPACE WING AND SELECTED SUBORDINATE UNITS)

45th Space Wing

Brig. General Jimmey R. Morrell	12 Nov 1991 – 29 Jun 1993
Brig. General Robert S. Dickman	30 Jun 1993 – 23 Jan 1995
Brig. General Donald G. Cook	24 Jan 1995 – 27 Aug 1995
Brig. General Robert C. Hinson	28 Aug 1995 – 26 Mar 1997
Brig. General F. Randall Starbuck	27 Mar 1997 – 19 Aug 1999
Brig. General Donald P. Pettit	20 Aug 1999 – 6 Jun 2002
Brig. General J. Gregory Pavlovich	7 Jun 2002 – 25 Aug 2004
Brig. General Mark H. Owen	26 Aug 2004 – 20 Jun 2006
Brig. General Susan J. Helms	21 Jun 2006 – 27 Oct 2008
Brig. General Edward L. Bolton, Jr.	28 Oct 2008 –

45th Operations Group

Colonel James N. Posey	12 Nov 1991 – 30 Jan 1992
Colonel Michael R. Spence	31 Jan 1992 – 19 Aug 1993
Colonel Glenn C. Waltman	20 Aug 1993 – 27 Apr 1995
Colonel Gary R. Harmon	28 Apr 1995 – 8 Jun 1997
Colonel Philip G. Benjamin, II	9 Jun 1997 – 23 May 1999
Colonel Darphaus L. Mitchell	24 May 1999 – 10 Jun 2001
Colonel Cameron S. Bowser	11 Jun 2001 – 6 Mar 2003
Colonel Gregory M. Billman	7 Mar 2003 – 28 Jun 2005

45th Operations Group

(Continued)

Colonel David D. Thompson	29 Jun 2005 – 11 Jul 2007
Colonel Bernard J. Gruber	12 Jul 2007 – 20 May 2009
Colonel James P. Ross	21 May 2009 –

45th Launch Group

Colonel Michael T. Baker	1 Dec 2003 – 22 Jul 2004
Colonel Samuel A. Greaves	23 Jul 2004 – 11 Jul 2006
Colonel Scott A. Henderson	12 Jul 2006 –

1st Space Launch Squadron

Note: No G-Series Orders or organizational charts mention the names of any 1 SLS Commanders prior to 12 November 1991. 45th Space Wing Special Order G-15 (dated 19 December 1991) confirms that Lt. Colonel Randolph M. Moyer was appointed the 1 SLS Commander as of 12 November 1991.

Lt. Colonel Randolph M. Moyer	12 Nov 1991 - 2 Aug 1993
Lt. Colonel Joseph Wysocki	3 Aug 1993 – 24 Nov 1994
Major Frank P. Arena	25 Nov 1994 – 20 Feb 1995
Lt. Colonel Doug D. Nowak	21 Feb 1995 – 30 Jul 1996
Major (later Lt. Colonel) Mark E. Dowhan	31 Jul 1996 – 6 Jul 1998
Lt. Colonel Blaise G. Kordell	7 Jul 1998 – 5 Jul 2000
Lt. Colonel David J. Buck	6 Jul 2000 – 30 Jun 2002
Lt. Colonel Brad T. Broemmel	1 Jul 2002 – 27 Jun 2004
Lt. Colonel Lavanson C. Coffey, III	28 Jun 2004 – 17 May 2006
Lt. Colonel Myron K. Fortson	18 May 2006 – 6 Jul 2008
Lt. Colonel Erik C. Bowman	7 Jul 2008 – to 1 SLS Inactivation

45th Spacecraft Operations Squadron

Lt. Colonel Ivory J. Morris	12 Nov 1991 – 7 Oct 1993
Lt. Colonel Larry D. James	8 Oct 1993 – 13 May 1994

Note: The 45th Spacecraft Operations Squadron was inactivated on 13 May 1994.

45th Operations Support Squadron

Lt. Colonel Heinz L. Butner	12 Nov 1991 – 31 Aug 1992
Lt. Colonel Timothy A. Roberts	1 Sep 1992 – 1 Sep 1994
Lt. Colonel James A. Schoeck	2 Sep 1994 – 30 May 1996
Lt. Colonel Dennis F. Hilley	31 May 1996 – 8 Jul 1998
Lt. Colonel David P. Trottier	9 Jul 1998 – 8 Jul 1999
Lt. Colonel Michael F. Kloskin	9 Jul 1999 – 12 Apr 2001
Lt. Colonel Kevin P. Karol	13 Apr 2001 – 15 Dec 2002
Lt. Colonel Kurt D. Hall	16 Dec 2002 – 12 Jul 2004
Lt. Colonel Christopher J. Kinnan	13 Jul 2004 – 24 Jul 2006
Lt. Colonel David G. Wilsey	25 Jul 2006 – 25 Sep 2008
Lt. Colonel John W. Giles, Jr.	26 Sep 2008 –

45th Range Squadron

Lt. Colonel Warren M. Greene	12 Nov 1991 – 11 Aug 1992
Lt. Colonel Elizabeth A. Enas	12 Aug 1992 – 13 Jul 1994
Lt. Colonel Richard H. Reynolds	14 Jul 1994 – 23 Mar 1995
Major Eric M. Mosby	24 Mar 1995 – 16 Jul 1995
Lt. Colonel David J. Thompson	17 Jul 1995 – 2 Mar 1997
Lt. Colonel Samuel A. Liburdi	3 Mar 1997 – 28 Mar 1999
Lt. Colonel Wayne L. Thompson	29 Mar 1999 – 1 Apr 2001
Lt. Colonel Andre L. Lovett	2 Apr 2001 – 9 Jan 2003

45th Range Squadron

(Continued)

Lt. Colonel Cynthia J. Grey   10 Jan 2003 – 1 Dec 2003

Note: The 45th Range Squadron was inactivated on 1 December 2003.

1st Range Operations Squadron

Lt. Colonel Cynthia J. Grey   1 Dec 2003 – 7 Jul 2004

Lt. Colonel Denette L. Sleeth   8 Jul 2004 – 13 Jul 2006

Lt. Colonel Peter B. Sterns   14 Jul 2006 – 15 May 2008

Lt. Colonel Brandt D. Nickell   16 May 2008 –

45th Launch Support Squadron

Lt. Colonel Scott L. Traxler   30 Jun 2005 – 16 May 2007

Lt. Colonel John W. Wagner   17 May 2007 – 30 Apr 2009

Lt. Colonel Erik C. Bowman   4 May 2009 –

APPENDIX B

YEAR-END STRENGTH FIGURES FOR THE 1 SLS AND RELATED UNITS

YEAR	UNIT	OFFICERS	ENLISTED	CIVILIANS	TOTAL
1990	DCS Operations	3	9	1	13
	1 SLS	20	13	4	37
	Ops Res Mgmt	4	14	25	43
	Payload Ops	24	18	11	53
1991	45 OG	5	3	3	11
	1 SLS	16	14	4	34
	45 SPOS	32	20	10	62
	45 OSS	13	100	11	124
	45 RANS	24	3	29	56
1992	45 OG	5	1	2	8
	1 SLS	16	12	4	32
	45 SPOS	24	23	7	54
	45 OSS	11	80	12	103
	45 RANS	15	4	22	41
1993	45 OG	5	1	2	8
	1 SLS	16	13	4	33
	45 SPOS	24	25	7	56
	45 OSS	11	78	12	101
	45 RANS	15	5	22	42
1994	45 OG	8	5	4	17
Note: 45 SPOS was inactivated on 13 May 1994	1 SLS	21	19	5	45
	45 SPOS	24	25	7	56
	45 OSS	13	78	12	103
	45 RANS	13	8	22	43
1995	45 OG	7	5	5	17
	1 SLS	31	20	5	56
	45 OSS	11	94	12	117
	45 RANS	17	8	27	52
1996	45 OG	9	6	4	19
	1 SLS	23	17	7	47
	45 OSS	13	85	15	113
	45 RANS	18	8	29	55
1997	45 OG	11	4	3	18
	1 SLS	18	20	5	43
	45 OSS	11	74	14	99
	45 RANS	15	9	26	50
1998	45 OG	10	5	3	18
	1 SLS	25	24	5	54
	45 OSS	10	56	10	76
	45 RANS	18	12	27	57
1999	45 OG	6	5	3	14
	1 SLS	19	25	5	49
	45 OSS	14	41	7	62
	45 RANS	19	17	25	61

YEAR	UNIT	OFFICERS	ENLISTED	CIVILIANS	TOTAL
2000	45 OG	10	8	4	22
	1 SLS	16	22	4	42
	45 OSS	14	37	5	56
	45 RANS	19	13	29	61
2001	45 OG	12	15	4	31
	1 SLS	17	20	4	41
	45 OSS	10	33	5	48
	45 RANS	20	14	28	62
2002	45 OG	9	7	4	20
	1 SLS	18	23	4	45
	45 OSS	11	37	6	54
	45 RANS	19	15	27	61
2003	45 OG	7	7	4	18
Note: 45 RANS was inactivated as 45 LCG and 1 ROPS were activated on 1 Dec 2003.	1 SLS	19	22	3	44
	45 OSS	16	49	5	70
	45 RANS	14	17	28	59
2004	45 OG	8	9	5	22
	45 LCG	3	4	2	9
	1 SLS	24	26	4	54
	45 OSS	11	39	5	55
	1 ROPS	18	13	29	60
2005	45 OG	10	6	3	19
Note: 45 LCSS was activated on 30 Jun 2005.	45 LCG	3	4	2	9
	1 SLS	24	18	4	46
	45 OSS	16	39	6	61
	1 ROPS	19	9	30	58
	45 LCSS	26	23	2	51
2006	45 OG	9	8	3	20
	45 LCG	3	5	2	10
	1 SLS	22	19	4	45
	45 OSS	16	30	11	57
	1 ROPS	22	10	31	63
	45 LCSS	26	30	6	62
2007	45 OG	22	10	8	40
Note: 45 OG numbers rose when HSFS was activated under 45 OG in April 2007.	45 LCG	4	4	2	10
	1 SLS	21	17	3	41
	45 OSS	11	19	19	49
	1 ROPS	18	6	30	54
	45 LCSS	19	29	4	52
2008	45 OG	21	14	10	45
	45 LCG	3	7	2	12
	1 SLS	14	13	3	30
	45 OSS	15	16	20	51
	1 ROPS	25	4	20	49
	45 LCSS	29	25	4	58

APPENDIX C

DELTA II & III LAUNCH SYNOPSIS

14 February 1989 – 24 Mar 2009

Date	Vehicle	Site	Payload
14 February 1989	DELTA II	Pad 17A	NAVSTAR GPS II-1 – 1ST GLOBAL POSITIONING SYSTEM (GPS) SATELLITE LAUNCH FROM THE EASTERN RANGE
10 June 1989	DELTA II	Pad 17A	NAVSTAR GPS II-2
18 August 1989	DELTA II	Pad 17A	NAVSTAR GPS II-3
21 October 1989	DELTA II	Pad 17A	NAVSTAR GPS II-4
11 December 1989	DELTA II	Pad 17A	NAVSTAR GPS II-5
24 January 1990	DELTA II	Pad 17A	NAVSTAR GPS II-6
14 February 1990	DELTA II	Pad 17B	LOSAT L & R
26 March 1990	DELTA II	Pad 17A	NAVSTAR GPS II-7
13 April 1990	DELTA II	Pad 17B	PALAPA B2R
1 June 1990	DELTA II	Pad 17A	ROSAT
2 August 1990	DELTA II	Pad 17A	NAVSTAR GPS II-8
18 August 1990	DELTA II	Pad 17B	BSB-R2
1 October 1990	DELTA II	Pad 17A	NAVSTAR GPS II-9
30 October 1990	DELTA II	Pad 17B	INMARSAT-2 F-1
26 November 1990	DELTA II	Pad 17A	NAVSTAR GPS II-10
8 January 1991	DELTA II	Pad 17B	NATO IV-A
8 March 1991	DELTA II	Pad 17B	INMARSAT-2 F-2
13 April 1991	DELTA II	Pad 17B	ASC-2
29 May 1991	DELTA II	Pad 17B	AURORA II
4 July 1991	DELTA II	Pad 17A	NAVSTAR II-11/LOSAT-X
23 February 1992	DELTA II	Pad 17B	NAVSTAR II-12
10 April 1992	DELTA II	Pad 17B	NAVSTAR II-13
14 May 1992	DELTA II	Pad 17B	PALAPA B4
7 June 1992	DELTA II	Pad 17A	EUVE
7 July 1992	DELTA II	Pad 17B	NAVSTAR II-14

Date	Vehicle	Site	Payload
24 July 1992	DELTA II	Pad 17A	GEOTAIL/DUVE
31 August 1992	DELTA II	Pad 17B	SATCOM C4
9 September 1992	DELTA II	Pad 17A	NAVSTAR II-15
12 October 1992	DELTA II	Pad 17B	DFS KOPERNIKUS 3
22 November 1992	DELTA II	Pad 17A	NAVSTAR II-16
18 December 1992	DELTA II	Pad 17B	NAVSTAR II-17
3 February 1993	DELTA II	Pad 17A	NAVSTAR GPS II-18
30 March 1993	DELTA II	Pad 17A	NAVSTAR GPS II-19
13 May 1993	DELTA II	Pad 17A	NAVSTAR GPS II-20
26 June 1993	DELTA II	Pad 17A	NAVSTAR GPS II-21
30 August 1993	DELTA II	Pad 17A	NAVSTAR GPS II-22
26 October 1993	DELTA II	Pad 17A	NAVSTAR GPS II-23
8 December 1993	DELTA II	Pad 17A	NATO IV-B
19 February 1994	DELTA II	Pad 17B	GALAXY I-R
10 March 1994	DELTA II	Pad 17A	NAVSTAR GPS II-24 & SEDS-2
1 November 1994	DELTA II	Pad 17B	WIND SPACECRAFT
5 August 1995	DELTA II	Pad 17B	KOREASAT-1
30 December 1995	DELTA II	Pad 17A	ROSSI X-RAY TIMING EXPLORER
14 January 1996	DELTA II	Pad 17B	KOREASAT-2
17 February 1996	DELTA II	Pad 17B	NEAR EARTH ASTEROID RENDEZVOUS (NEAR)
28 March 1996	DELTA II	Pad 17B	NAVSTAR GPS II-25
24 May 1996	DELTA II	Pad 17B	GALAXY IX
16 July 1996	DELTA II	Pad 17A	NAVSTAR GPS II-26
12 September 1996	DELTA II	Pad 17A	NAVSTAR GPS II-27
7 November 1996	DELTA II	Pad 17A	MARS GLOBAL SURVEYOR
4 December 1996	DELTA II	Pad 17B	MARS PATHFINDER
17 January 1997	DELTA II	Pad 17A	NAVSTAR GPS IIR-1
20 May 1997	DELTA II	Pad 17A	THOR II
23 July 1997	DELTA II	Pad 17A	NAVSTAR GPS IIR-2
25 August 1997	DELTA II	Pad 17A	ADVANCED COMPOSITION EXPLORER
6 November 1997	DELTA II	Pad 17A	NAVSTAR GPS II-28

Date	Vehicle	Site	Payload
10 January 1998	DELTA II	Pad 17B	SKYNET 4D
14 February 1998	DELTA II	Pad 17A	GLOBALSTAR-1
24 April 1998	DELTA II	Pad 17A	GLOBALSTAR-2
10 June 1998	DELTA II	Pad 17A	THOR III
27 August 1998	DELTA III	Pad 17B	GALAXY X
24 October 1998	DELTA II	Pad 17A	DEEP SPACE 1
22 November 1998	DELTA II	Pad 17B	BONUM-1
11 December 1998	DELTA II	Pad 17A	MARS CLIMATE ORBITER
3 January 1999	DELTA II	Pad 17B	MARS POLAR LANDER
7 February 1999	DELTA II	Pad 17A	STARDUST (COMET PROBE)
5 May 1999	DELTA III	Pad 17B	ORION-3
10 June 1999	DELTA II	Pad 17B	GLOBALSTAR-3
24 June 1999	DELTA II	Pad 17A	FUSE (SPACE OBSERVATORY)
10 July 1999	DELTA II	Pad 17B	GLOBALSTAR-4
25 July 1999	DELTA II	Pad 17A	GLOBALSTAR-5
17 August 1999	DELTA II	Pad 17B	GLOBALSTAR-6
7 October 1999	DELTA II	Pad 17A	NAVSTAR GPS IIR-3
8 February 2000	DELTA II	Pad 17B	GLOBALSTAR-7
11 May 2000	DELTA II	Pad 17A	NAVSTAR GPS IIR-4
16 July 2000	DELTA II	Pad 17A	NAVSTAR GPS IIR-5
23 August 2000	DELTA III	Pad 17B	DM-F3 (DUMMY PAYLOAD)
10 November 2000	DELTA II	Pad 17A	NAVSTAR GPS IIR-6
30 January 2001	DELTA II	Pad 17A	NAVSTAR GPS IIR-7
7 April 2001	DELTA II	Pad 17A	MARS ODYSSEY
18 May 2001	DELTA II	Pad 17B	GEOLITE (TECHNOLOGY EXPERIMENT)
30 June 2001	DELTA II	Pad 17B	MICROWAVE ANISOTROPY PROBE
8 August 2001	DELTA II	Pad 17A	GENESIS SPACECRAFT
3 July 2002	DELTA II	Pad 17A	COMET NUCLEUS TOUR

Date	Vehicle	Site	Payload
29 January 2003	DELTA II	Pad 17B	NAVSTAR GPS IIR-8
31 March 2003	DELTA II	Pad 17A	NAVSTAR GPS IIR-9
10 June 2003	DELTA II	Pad 17A	MARS EXPLORATION ROVER-A
8 July 2003	DELTA II	Pad 17B	MARS EXPLORATION ROVER-B
25 August 2003	DELTA II	Pad 17B	SPACE INFRARED TELESCOPE FACILITY
21 December 2003	DELTA II	Pad 17A	NAVSTAR GPS IIR-10
20 March 2004	DELTA II	Pad 17B	NAVSTAR GPS IIR-11
23 June 2004	DELTA II	Pad 17B	NAVSTAR GPS IIR-12
3 August 2004	DELTA II	Pad 17B	MESSENGER
6 November 2004	DELTA II	Pad 17B	NAVSTAR GPS IIR-13
20 November 2004	DELTA II	Pad 17A	SWIFT SPACECRAFT
12 January 2005	DELTA II	Pad 17B	DEEP IMPACT
26 September 2005	DELTA II	Pad 17A	NAVSTAR GPS IIR-14 (M)
21 June 2006	DELTA II	Pad 17A	MICROSATELLITE TECHNOLOGY EXPERIMENT (MITEX)
25 September 2006	DELTA II	Pad 17A	NAVSTAR GPS IIR-15 (M)
26 October 2006	DELTA II	Pad 17B	STEREO (SOLAR OBSERVATORY)
17 November 2006	DELTA II	Pad 17A	NAVSTAR GPS IIR-16 (M)
17 February 2007	DELTA II	Pad 17B	THEMIS (MAGNETIC PROBES)
4 August 2007	DELTA II	Pad 17A	PHOENIX (MARS PROBE)
27 September 2007	DELTA II	Pad 17B	DAWN (ASTEROID PROBE)
17 October 2007	DELTA II	Pad 17A	NAVSTAR GPS IIR-17 (M)
20 December 2007	DELTA II	Pad 17A	NAVSTAR GPS IIR-18 (M)
15 March 2008	DELTA II	Pad 17A	NAVSTAR GPS IIR-19 (M)
11 June 2008	DELTA II	Pad 17B	GLAST (SPACE TELESCOPE)
7 Mar 2009	DELTA II	Pad 17B	KEPLER SPACECRAFT
24 Mar 2009	DELTA II	Pad 17A	NAVSTAR GPS IIR-20 (M)

DELTA II NAVSTAR GPS MISSION PATCHES